THE PHILOSOPHER'S STONE

*Michio Kushi's Guide to Alchemy,
Transmutation & the New Science*

MICHIO KUSHI
EDWARD ESKO

With Contributions by Alex Jack,
Mahadeva Srinivasan & Matthias Grabiak

IMI Press
Lenox, Mass.

THE PHILOSOPHER'S STONE
Michio Kushi's Guide to Alchemy, Transmutation & the New Science

Published by IMI Press
An Imprint of the International Macrobiotic Institute (IMI)
P.O. Box 2051
Lenox, MA 01240

InternationalMacrobioticInstitute.com
(413) 446-2620
edwardesko@gmail.com

One Peaceful World Press Edition: December 1994
IMI Press Edition: August 2020

Note to the Reader:
The experiments and procedures described in this book are for educational purposes only.
Please do not attempt to duplicate without the guidance and participation of trained and
qualified engineering, electrical, chemical, nuclear, or materials professionals. Please observe
all health and safety precautions, including those spelled out in the Materials Safety Data
Sheet (MSDS) for specific elements, as well as the advice of expert professionals, the use of
protective eyewear, and the proper disposal of hazardous chemicals.

We are probably on the eve of a great change in physics.
C. LOUIS KERVRAN

CONTENTS

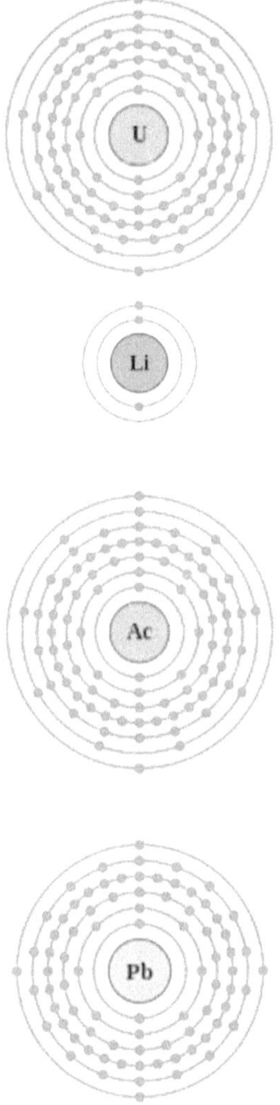

The low energy fission of ^{235}U into ^{228}Ac could shorten the half life of uranium-235 from 700 million years to 1.9 years

Atomic Number and Atomic Mass*

When you study the periodic table, the first thing that you may notice is the number that lies above the symbol. This number is known as the **atomic number**, which identifies the number of protons in the nucleus of ALL atoms in a given element. The symbol for the atomic number is designated with the letter **Z**. For example, the atomic number (z) for sodium (Na) is 11. That means that all sodium atoms have 11 protons. If you change the atomic number to 12, you are no longer dealing with sodium atoms, but magnesium atoms. Hence, the atomic number defines the element in question.

Recall that the nuclei of most atoms contain neutrons as well as protons. Unlike protons, the number of neutrons is not absolutely fixed for most elements. Atoms that have the same number of protons, and hence the same atomic number, but different numbers of neutrons are called **isotopes**. All isotopes of an element have the same number of protons and electrons, which means they exhibit the same chemistry. Because different isotopes of the same element haves different number of neutrons, each of these isotopes will have a different **mass number(A)**, which is the sum of the number of protons and the number of neutrons in the nucleus of an atom.

Mass Number(A) = Number of Protons + Number of Neutrons

The element carbon (C) has an atomic number of 6, which means that all neutral carbon atoms contain 6 protons and 6 electrons. In a typical sample of carbon-containing material, 98.89% of the carbon atoms also contain 6 neutrons, so each has a mass number of 12. An isotope of any element can be uniquely represented as $^{A}_{Z}X$, where X is the atomic symbol of the element, A is the mass number and Z is the atomic number. The isotope of carbon that has 6 neutrons is therefore $^{12}_{6}C$. The subscript indicating the atomic number is actually redundant because the atomic symbol already uniquely specifies Z. Consequently, it is more often written as ^{12}C, which is read as "carbon-12." Nevertheless, the value of Z is commonly included in the notation for nuclear reactions because these reactions involve changes in Z.

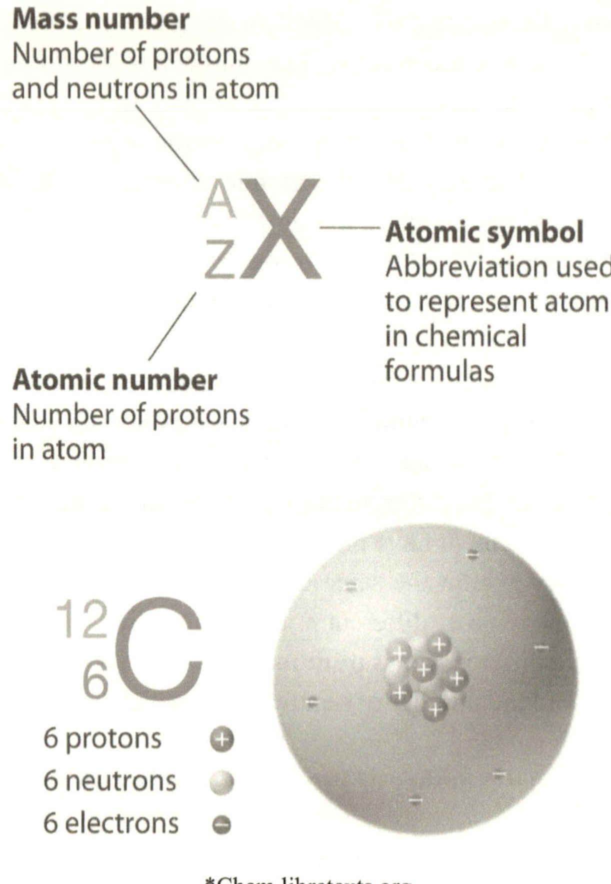

*Chem.libretexts.org

Preface to the 2020 Edition

It is with great pleasure that we present this new updated edition of *The Philosopher's Stone*. Much has taken place in the field since it was published in 1994, both in our own research and thinking, and in parallel developments around the globe.

This new edition combines Michio Kushi's original volume with newer material. Michio's original text appears as Part I. Part II includes my articles and essays, many of which were published in *Infinite Energy*, the journal of new energy and science, and compiled in the book, *Cool Fusion* (2012), authored with Alex Jack. More recent essays are included in Parts III and IV, and are borrowed from *IE* articles and the books, *Corking the Nuclear Genie* (2013) and *In Search of Nanonovae* (2018.)

Michio Kushi was actively involved in our research until his passing in 2014. My Quantum Rabbit partners, Alex Jack and Woody Johnson, and I would meet with him often to describe our results and to seek advice and counsel. Michio was excited and happy with our experiments as they repeatedly validated his earlier work and predictions. In 2011, we hosted a meeting with Mahadeva Srinivasan, former director of the physics group at the Bhabha Atomic Research Center in India. Earlier, Michio had suggested that we focus our efforts on critical environmental concerns, especially the growing problem of nuclear waste. We met Dr. Srinivasan to discuss this issue and the use of low energy transmutation in solving problems of economy, ecology, and natural resources.

Since that time, I have been networking with colleagues around the world who are working on their own unique approaches to these problems, including, in addition to Dr. Srinivasan, Vladimir Visotskii in Ukraine and George Egly in Hungary. These pioneers are conducting work on low energy transmutation and have agreed to serve as consultants to the Quantum Research Institute (QRI), our advanced research and study group.

Our most robust experimental results were achieved at the Nashua and Owl's Head labs from 2008-2013. They are summarized in this book. Following this peak, experimental results were less definitive. We discovered that this was due to aging and underperforming equipment, especially the vacuum system that was assembled in 2005. We are currently in the process of redesigning and upgrading this essential hardware so as to advance our research in the coming decade.

Part of the problem in advancing low energy transmutation is that accepting its foundational premise requires a quantum leap in thinking, or at least an open mind. It is doubtful whether Big Science and Big Physics, as they now exist, are ready to make that leap. One of the simplest tenets of transmutation is the fact that elementary particles—atoms, electrons, protons, and neutrons—are not identical. They are like snowflakes. They share a basic configuration, but each is completely unique. No two atoms are exactly the same, thus there exists some degree of difference, and hence polarity and attraction, between them. However, even progressive thinkers who acknowledge the possibility of unconventional approaches have a problem with this notion. Here is a recent email exchange with a colleague in the cold fusion community:

Quick question: What do you think of "non-identity?" For example no two protons, electrons, neutrons, atoms, etc. are exactly alike. Each is completely unique. Do you consider this to be an "unproven theory?"

His reply was surprising:

I have little to say about "non-identity" other than my physics courses, and what I have learned since; do not attribute anything to particles that would distinguish them.

At least he was forthright and honest. We're clearly dealing with a one-size-fits all view of the world, including the concept of elemental particles that make up our physical world. No wonder the language of transmutation falls on deaf ears. It is the language of the future and thus difficult, if not impossible, for those trained in modern concepts to comprehend. It is a question of change and flexibility versus a static and inflexible mindset. Although published in the leading new energy journal, and in our books, the Quantum Rabbit research, as dramatic as it is, was not so much disputed or challenged as it was simply *ignored*. The lack of curiosity on the part of the scientific community seems quite frankly, to be rather unscientific.

Another problem that we face in moving things forward is the phenomenon known as "not in my backyard," or "NIMBY." The instinct for survival and self-preservation behind this phenomenon is of course essential for health and wellbeing. It is fine as far as it goes. However, if this thinking leads to a sense of complacency and a lack of desire to find genuine solutions, it can be unhelpful. My recent correspondence with an anti-nuclear activist in New Mexico illustrates the point.

My friend in New Mexico is helping lead a coalition of concerned citizens, including tribal leaders, opposed to the siting of a temporary storage facility for U.S. nuclear waste in his state. Holtec International, a New Jersey corporation that specializes in storing spent nuclear fuel, recently purchased 1,000 acres of desert in southeastern New Mexico for a "consolidated interim storage facility." The plan is to house 120,000 metric tons of nuclear waste over 40 years, at least until a more "permanent" facility is found. Ironically, New Mexico was the site of the Manhattan Project, the WWII project that led to the creation of the atomic bomb and to nuclear power.

The Holtec project would involve shipping highly radioactive waste by rail and truck from all over the U.S. with shipping routes crossing populated areas including New Mexico tribal lands. A physicist friend mentioned to him that transmutation might offer a long-term solution for nuclear waste and that he should contact me. I sent copies of our paper, "Low Energy Transmutation of Nuclear Waste" for distribution at a recent conference on the issue and offered to work with local activists in designing research that could deal with nuclear waste at the local level and eliminate the need for a nationwide storage facility. Once again, however, the language of the future failed to resonate. My activist friend recently emailed:

One engineer told me not to mention transmutation in my talks until there is verifiable proof. Now I am focused on robust storage in situ until there is a viable solution.

As I point out in my paper, on site storage is not a solution. Nor is collective storage such as that proposed for New Mexico, or various other proposals for "permanent" storage. The only "solution" for manmade nuclear waste is to eliminate nuclear waste. Period. Any other proposed "solutions" are nothing more than burrowing one's head in the sand and kicking the can down the road for future generations to deal with.

Another problem in moving forward is caused by the fragmentation of research into an unaccountable number of specialties, each with its own agenda and need for funds. The modern paradigm is clearly big business. On the MIT website, the sources of funding for 2019 research are itemized. MIT's budget for that year totaled $773.9 million. Of that amount private industry funded roughly 22%, the Pentagon, 18%, the Department of Energy, 9%, NASA, 4%, and other federal agencies, 2%. Without government and corporate support, Quantum Rabbit's yearly budget was a miniscule fraction of MIT's. Clearly, in this environment it is easy for research on low energy transmutation to get lost in the shuffle and drowned out by the noise. With no unified consensus, modern science continues to wander in the darkness and down the same old path.

However, a unified consensus is emerging in spite of these challenges. In 2019 I made contact with Wal Thornill, the Australian physicist noted for his work on the electric universe. Wal had produced a video on the Safire Project, a series of experiments designed to explore the role of electricity in stellar and planetary phenomena. Like the Quantum Rabbit experiments, Safire research takes place in an electrified vacuum chamber. As Wal reported, one of the results of these tests was the appearance of anomalous elements thought to be the result of low energy transmutation. I sent Wal an email describing our work and a PDF of our book, *Cool Fusion*. Wal replied:

On first glance your book looks like a fine production, which I'm eager to read. Nuclear remediation is a critical problem, which I'm sure can be solved in a relatively low energy plasma environment. So I will be pleased to refer to your work appropriately in future.

Wal and his associates are developing a cosmology that dovetails with that of low energy transmutation. The universe is comprised of endless streams or currents of energy, referred to as plasma, which continually appear, manifest, and disappear back into the world of energy. These currents resemble fractal branches, continually branching and dividing, as Michio Kushi explains in this book, into the complementary opposite movements known as yin and yang. Also, as he points out, these endlessly appearing branches take the form of expanding and contracting spirals. On the macroscopic level, they produce galaxies, star systems, and planets. On the microscopic level, they condense into subatomic particles, atoms, and the elements of our material world.

This process is peaceful, never ending, and constantly changing. The cosmology of change negates limited conceptions such as the big bang and thermonuclear sun in favor of an endless process of change and transformation. As our friends at the electric universe point out, electromagnetism is at the core of material existence. It is the force that obeys the laws of yin and yang and out of which atoms, elements, and life come into being. Low energy transmutation holds great promise, both as a practical tool for solving critical environmental problems, and as a crucial step in the development of a unified cosmology that will benefit humanity as a whole.

Edward Esko
July 2020

PART I

The Philosopher's Stone

Much progress has taken place in establishing firmly
the occurrence of different types of transmutation
reactions in a wide variety of configurations.
MAHADEVA SRINIVASAN, BHABHA ATOMIC
RESEARCH CENTER (RETIRED)

FOREWORD

Making gold and other precious metals from inexpensive, common materials has intrigued the most brilliant minds since the dawn of history. In both East and West, alchemists have labored over furnaces and crucibles in an effort to uncover the secrets of transmutation. Isaac Newton, who discovered the universal law of gravitation, saw his work on celestial mechanics as a diversion from his more fundamental studies on alchemy and the underlying vibrational harmony of all matter.

In our own era, macrobiotic educators George Ohsawa and Michio Kushi revived this age-old quest about thirty-five years ago in a series of experiments in Tokyo and Cambridge, Mass. Using the unifying principle of yin and yang, they discovered that the Periodic Table of Elements was actually a dynamic spiral, not a fixed grid, and that elements were constantly changing into each other at normal temperatures. In the world of chemistry, this discovery amounts to the kind of revolutionary insight that transformed the other sciences. Galileo, Newton, and the 17th century natural philosophers proved that the earth was not the fixed center of the cosmos, but a small planet in a constantly changing universe. In the 19th century, Charles Darwin broke with millennia of dogma and showed that species were not immutable but evolved, or descended, from common ancestors.

The modem belief that chemical elements are static and cannot be transmuted except by atomic fission, hydrogen fusion, or other high tech methods has led to the development of the Manhattan Project, the Supercollider Project, and other complex, destructive, costly campaigns to unlock the secrets of the atom. As George Ohsawa, Michio Kushi, and Louis Kervran—a French scientist who initiated some remarkable experiments in biological transmutation in the 1950s and 1960s—show, a practically limitless source of energy is available using common, simple, safe elements and compounds.

Much practical investigation remains to be done to actually develop this principle and mass-produce iron, steel, platinum, gold, silver, and other metals. During the Cold War, U.S. Army (and presumably Soviet) scientists investigated and confirmed the early experiments described in this book. Recently, scientists in Texas have begun tests based on George Ohsawa and Michio Kushi's teachings and also reported positive results.

Beginning this year, Michio plans to meet with representatives of some of the world's major industries to interest them in developing this method. His purpose is to launch a New Industrial Revolution that would make hitherto scarce resources available to everyone at an affordable price, preserve the earth's natural environment (which is rapidly declining as a result of industrialization), and drive down the international competition for precious materials that is a major cause of poverty, pollution, war and conflict. While this may take many years to realize, the world as a whole will ultimately become more healthful and peaceful. This book is the opening manifesto in the New Industrial Revolution that will one day circle the planet and inaugurate a new era of peace and freedom.

Alex Jack
Becket, Mass.
February 4, 1994

Introduction

The study of atomic transmutation has intrigued me since the beginning of my macrobiotic practice over twenty years ago. Much of the early macrobiotic literature contained references to the work of Louis Kervran, George Ohsawa, and others in the field of transmutation. Kervran, a French biochemist, discovered the transmutation of sodium into potassium in French workers in the Sahara. His findings were published in the book, *Biological Transmutations*. Mr. Ohsawa worked with Kervran and devoted the latter part of his life to proving transmutation in the laboratory. As you will discover in this book, transmutation offers proof of the mutability of the material world, and is at the core of the macrobiotic philosophy of change.

When I visited Prague several years ago, friends took me to a section of the city called "Alchemists' Row." It is a narrow street on either side of which are curious tiny houses. Our guide explained that these houses were where medieval alchemists had lived and conducted their experiments. Prague was one of the centers of medieval European alchemy. The strange little houses on Alchemists' Row now serve as boutiques and gift shops for tourists.

Over the years, Michio Kushi has given lectures on transmutation, and transcripts of these lectures were published in his seminar reports. When I edited Michio's book, *Other Dimensions: Exploring the Unexplained* (Avery, 1992), I included a chapter on transmutation. I have also introduced the principles of transmutation in lectures at the East West Foundation in Boston and later as a part of the Kushi Institute program in Becket.

Modem chemistry and physics were born in the laboratories of the medieval alchemists. However, in the twentieth century, scientific pursuit of alchemy took a destructive turn. In contrast to the peaceful, natural methods employed by ancient alchemists, modem researchers began to utilize violent and destructive methods to achieve transmutation. In 1920, Rutherford changed nitrogen into hydrogen and oxygen by bombarding nitrogen atoms with subatomic alpha particles. Ten years later, Ernest Lawrence invented a device called the cyclotron, or particle accelerator; in which atomic particles are accelerated with tremendously high energy and used to "smash" target atoms. In 1932, scientists discovered that neutrons could be used as "bullets" to smash atoms, and in 1939, the nucleus of uranium was bombarded with neutrons, causing it to "split" and release energy. In 1942, Enrico Fermi at the University of Chicago used this discovery to achieve a chain reaction, and soon afterward, Oppenheimer and other researchers in the Manhattan Project used these discoveries to build the first atomic bomb.

The discovery of peaceful, natural transmutation offers an alternative to these destructive methods. Atomic transmutation is taking place in nature and can be duplicated without having to attack and destroy atoms. The work of George Ohsawa, Louis Kervran, Michio Kushi, and other pioneers in peaceful, natural transmutation has shown that matter is not fixed and static, but dynamic and changing. The information in this book has the potential to eclipse the discovery of atomic energy, revolutionize science and technology, and open the door to a new era for humanity. If this knowledge is properly understood and applied, the age-old quest for the philosopher's stone will contribute to an era of peace and unprecedented prosperity. The transmutation of the atom may hold the key to the transmutation of society itself.

Edward Esko
January 1994

Two investigators, Kervran and Komaki, have been recently nominated for a joint Nobel Prize for their work involving experimental proof that elemental transmutations were occurring in life organisms. Elements which were definitely proven to have been transformed were sodium (to magnesium), potassium (to calcium), and manganese (to iron). Actually, observations have been made for almost 200 years that elemental transmutations were occurring, but little credence was given to them because they resembled alchemy—a relic of the Middle Ages.
1978 U.S. ARMY REPORT ON BIOLOGICAL TRANSMUTATIONS

1

MATTER AND ENERGY

Let us begin our study of alchemy and transmutation with a review of the basic laws of matter according to the unifying principle of yin and yang. There are four states of matter. The first state is a very hard one—solid. The second state is liquid. The third state is gas. And the fourth state is plasma. The first three are the states of matter. Plasma is the state that exists between matter and energy. It is the waves or vibrations surrounding the other states of matter. For example, you can see the waves that surround a flame when a cigarette lighter is lit. That is the plasma state of the gas, usually butane, which is being burned up.

The atomic density of a solid is more compact than that of a liquid or a gas. It disperses as the solid becomes a liquid, and the liquid becomes a gas. In the plasma state, the atomic structure is very loose. Electrons and protons are separating from each other and ultimately changing into energy. At that point, matter can no longer be identified. Beyond this state exists the world of pure energy or vibration. In religious or philosophical terms, that world is referred to as the world of spirit.

On the earth and throughout the universe, matter continually changes into energy. Energy continually changes into matter. Solids are very condensed. They are yang. In order to make them change into liquids, what do we need to apply? We need to apply yang, in the form of heat. Then the solid will melt and change into a liquid. If we keep applying heat, it will boil and change into a gas. More heat will cause the atoms to break apart into electrons and protons. Eventually, these preatomic particles will dissolve into energy.

If we apply yin—cold—then the opposite process takes place. Energy condenses into matter. When water vapor cools, for example, it condenses into liquid water. If it cools further, it becomes a solid, ice. But if we heat the ice, it will melt into a liquid. If we heat it further, we get a gas, steam. If we heat it still further, we will get plasma, and then finally it will disappear into the world of vibration. The stages of matter are presented below.

Stages of Matter and Energy

Heat

Spirit	Non-matter (vibration, waves, energy)
	4. Plasma (border between matter and energy)
Matter	3. Gas
	2. Liquid
	1.Solid

Cold

In terms of temperature, solids are very yin, very cold. What is a piece of chalk? Hard. Solid. Just like ice. Solid matter is hard and cold. When you grip metal, it is cold. Metal has no heat of its own. When you touch plastic, it is cold. When you touch your shoes, they are cold. All solids are frozen. The floor, the ceiling, the glass in the window; all are cold and frozen. In their gaseous state, elements are not frozen. They are active or energized. The state of the element depends upon its melting and boiling point. All elements melt into liquids and boil into gases at different temperatures. When the melting point is very low in comparison to room temperature, the element exists in the gaseous state. Is the structure of a gas more yin or more yang? Solids are like ice. Unless we apply heat, they do not melt. When viewed in terms of structure, solids are condensed, yang. Gases are diffused or yin. Here we can see that yin—cold—produces yang—contraction and solidity; while yang—heat—produces yin—expansion and diffusion.

2

ATOMIC STRUCTURE

In school we learned that the first atom is hydrogen with an atomic number of one, and an atomic weight of one. That means it has one electron and one proton. Is the electron yin or yang? More yin. It occupies a peripheral position in the atom and has a negative charge. The proton is yang. It is in the center and has a positive charge. Electrons and protons make balance with each other, and that is why atoms are electrically neutral.

What is the neutron? The neutron has no electric charge. No plus or minus. That means no yin or yang. However, can such a thing actually exist? In order to understand this, we must understand the structure of the atom. We learned the structure of the atom in high school: electrons revolving in fixed orbits around protons in the nucleus. But is that the true picture? Are electrons permanently locked into fixed orbits? If we agree that everything changes, then the distance between electrons and protons must be changing.

Everything moves in expanding or contracting spirals. The current model of the atom shows electrons moving in a static circle. But that is actually a hypothesis, meaning that it is a product of someone's imagination. No one actually saw whether or not atoms have this structure. According to the unifying principle, however, yin changes into yang, and yang changes into yin. Yin and yang attract each other. Electrons gradually migrate toward the center of the atom. Electrons change into protons. Protons are actually transformed electrons.

The neutron is the intermediate stage between the inward and outward spirals that give rise to atoms. According to the spiral model, electrons become protons, protons become neutrons, and neutrons become electrons. When these inward and outward spirals reach a state of balance or equilibrium, they become practically invisible. This state of balance is referred to as the "neutron." Actually, the neutron is not perfectly balanced; it has a very tiny charge that it is almost impossible to detect.

The electron has a very small mass and weight. The proton is very big. With such a difference in size, how do they make balance with each other? First, the electron's motion is big. The electron has an active, yang motion. The proton's motion is yin. It is rotating, but more slowly than the electron. The difference in motion compensates for the difference in mass between particles at the periphery and those at the center of the atom. In between the electron and the proton there are many other particles, also moving in a spiral pattern. These particles are gathering toward the center of the atom, toward the proton. So the proton becomes big.

Since we have given a name to the proton, we make the mistake of thinking it is a distinct entity that we can extract. But it is not like that. These small particles gather toward the center with enormous pressure and speed. And that total mass we call the proton. The proton is not one single mass, but a gathering of many tiny particles.

Moreover, the electron is not like a tiny billiard ball. The electron is a tightly coiled spiral made up of many tinier particles. Try not to visualize electrons, protons, and neutrons as discrete "parts" of the atom that can be lifted out and examined. All those things are nothing but small particles, moving in spirals, developing from one stage into another. Those stages have been named electron, proton, and neutron. And we have become confused by the use of this terminology.

Even the particles themselves, what are they? Are they tiny specks of matter? They are condensed spirals of energy. It appears that they are "particles" when we test the atom for their presence. But in reality there is nothing that can be called a "particle." Searching for the matter in the atom is like searching for a ghost. No one can find it. There is no substance in this universe, only change. Existence is equal to change or movement. Everything is change, and change is everything. The law of change covers the whole of existence. It is the law that governs the worlds of matter and energy.

3

ATOMIC ENERGY

Atoms represent the condensation of a tremendous amount of energy. The sudden conversion of matter to energy results in an atomic reaction. An example of this is the atomic bomb. The atomic bomb explodes fiercely, suddenly, with tremendous destructive power. In Hiroshima, for example, 360,000 people were killed instantly by the atomic blast. In Nagasaki, about half the population perished immediately. An atomic explosion causes lethal radiation to spread. The spread of radiation is a very yin phenomenon. If nuclear war had ever broken out between the United States and the Soviet Union, a short time after the detonation of an American warhead over Moscow, a radioactive cloud would have reached New York. Because of the extremely yin nature of radiation, nuclear war can never remain "limited." Destruction would take place on a planetary scale.

Throughout the world, nuclear power plants are used to supply energy. However, atomic power also generates a great deal of radioactive waste. Disposing of radioactive waste is one of the greatest environmental problems facing us today. If radioactivity is yin, how can we prevent it from spreading? We need to change it into its opposite. Extreme cold causes things to condense or solidify, and thus could help contain it. Radioactivity is yin. The application of yang—heat—will cause it to disperse more widely. Why was Hiroshima destroyed completely with only one atomic bomb? More yang conditions prevailed—it was a hot sunny day in August. But if the bomb had exploded on a cold winter day, the effect would not have been so devastating. In order to reduce the spread of radioactivity, therefore, we should immerse the radioactive waste in a very cold substance such as liquid nitrogen.

Yin and yang can help us solve many technical problems. Once you understand these principles, you can do many wonderful things. The elemental, atomic world will never cheat. It will never tell a lie. People cheat all the time. There is never an accurate result in an experiment dealing with people. But the world of matter is completely reliable. So if we apply the principles of yin and yang accurately in this domain, the results will be accurate and predictable.

4

THE SPIRAL OF THE ELEMENTS

When we studied the periodic table of the elements in school, we did not learn it according to the unifying principle or the order of the universe. The understanding of the elements is lacking in the largest, or if you will, the macrobiotic view. In the periodic table, how many sections are there from top to bottom? There are seven. That immediately makes you think of how commonly this number of divisions appears, for example, in the seven colors of the visible spectrum, the seven tones of the musical scale, the seven chakras, or energy centers, in the human body, and the seven days of the week.

We can also count the number of divisions from left to right. With the exception of the noble gases, there are twelve divisions. What do you imagine from this general construction—seven and twelve? This pattern is a reflection of the seven orbits of the spiral of materialization and the twelve divisions of that spiral. This pattern is universal: it is found throughout nature. We see the number twelve reflected in the twelve months of the year; the influence of the twelve constellations, and the twelve meridians of energy that run through the human body.

All things appear and disappear in the form of spirals. The world of matter, or the world of elements, from solid to plasma, is constructed in a spiral form. In arranging the elements in a spiral, the lighter elements, such as hydrogen and helium, are located at the periphery, or outermost orbits, while the heavier elements are found toward the center. The first orbit of the spiral contains hydrogen and helium. The second orbit consists of lithium, beryllium, boron, carbon, nitrogen, oxygen, fluorine, and neon. The third orbit begins with sodium. The fourth begins with potassium, and so on. If we check the elements that lie within the seventh orbit, such as uranium, we wee that they are radioactive.

Periodic Table of the Elements

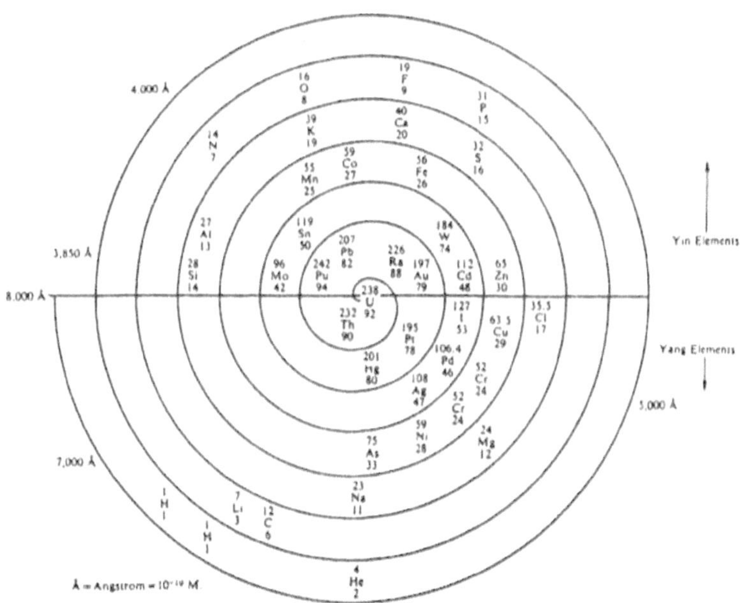

Spiral of the Elements

Now we can understand why these heavy elements are radioactive. The heavy elements at the center of the spiral are in the process of decomposing. After a certain period, uranium will decompose and turn into lead. Lead has the periodic number 82. It lies on the sixth orbit of the spiral. Uranium does not remain within the seventh orbit for very long; it eventually decays and returns to the sixth orbit.

Lighter elements become heavier elements. Then, at the extreme of heaviness, the elements become radioactive and return toward lighter elements. This tendency we can see clearly. The centripetal, yang force of the universe causes lighter elements to become heavier ones. The yin, centrifugal force of the universe causes heavier elements to become lighter ones in an endless cycle. Elements in the third, fourth, and fifth orbits are located in the middle region of the spiral. They are generally stable. All of the elements can be classified from yin to yang; each has a slightly different position on the spiral.

The fourth orbit lies in the middle of the spiral, in between the outermost and innermost orbits. It occupies the most balanced position. The elements that lie within the fourth orbit are iron, cobalt, and nickel. Among the elements, these have a unique property: they are magnetic. What is magnetism? Magnetism is attraction. But why does the attraction happen? Of all the matter in the universe, these three elements are at the center. All of the other elements gravitate toward that center. In modern thinking, we say that iron has magnetism. But that way of thinking is backwards. Everything in the universe is moving toward a state of balance. And in the material world, that state of balance exists in the form of these magnetic elements. Iron does not generate a magnetic field. If we could remove all the other elements from the universe, and leave only iron behind, then the iron would no longer "have" magnetism. Magnetism exists prior to iron. Iron does not have a magnetic character. Iron exists as the result of magnetism. Magnetism is the force in nature that causes all things to seek a state of balance or equilibrium.

There are several methods we can use to determine whether an element has a more yin or more yang character. One is to check the atomic weight. If the weight is heavier, the element is more yang. Yin elements have lighter atomic weights. A second method is to check the physical reaction, for example, to heat and pressure. Elements that resist heat and pressure are yang. Elements that are easily changed by heat and pressure are yin.

Metals and other elements that resist heat—which means they have high melting and boiling points—are yang; those with low melting and boiling points, such as gases, are yin. A third method is to check the chemical reaction. We know that oxygen is necessary to feed fire. Fire is yang, so therefore oxygen is a yin element. Then if oxygen is a yin element, and we know that it combines easily with carbon, that means carbon is a more yang element. All chemical combinations happen according to yin and yang. There are no exceptions to these simple laws.

A fourth method is to check the color. The color of an element is a reflection of its wavelength, and each element has its own distinctive color. Scientists use a spectroscope to analyze color and wavelength. Wavelengths are measured in angstrom units. The color green ranges from around 3,000 to 8,000 angstroms. By analyzing the wavelength of each element, their color can be determined, and the element positioned on the spiral according to its yin or yang nature. Elements that appear at the orange-yellow end of the spectrum are more yang; those appearing at the opposite, or green-blue end are more yin.

With these methods, we can construct a beautiful spiral chart of the elements; arranging them from yin to yang. This chart can replace the current periodic table, since it is more practical, dynamic, and realistic. The elements are actually stages in a continuum. They are not separate and distinct. They do not exist independently of one another, but are part of a spiral process of change. Even though it may take billions of years, a hydrogen atom will eventually change into heavier and heavier elements. And once it reaches its heaviest state, it begins to change back to the lightest state—the hydrogen atom. The movement from expansion to contraction and from contraction back to expansion occurs everywhere in nature. The behavior of atoms and elements is but a reflection of the endless order of change.

5

NEW ALCHEMY

Modern science has been in existence for only several centuries. Before that was the age of alchemy. In Europe, the work of the medieval alchemists gave birth to modern physics and chemistry. Alchemy may also have been widely practiced in the ancient world. More than forty years ago, a large tomb was unearthed in northern China. It is about 3,000 years old, and is thought to have been built for a prominent general. A small, yellow-gold box, made of some type of metal, was found in the tomb. It was analyzed at a university in Beijing and discovered to contain more than 70 percent aluminum. This discovery is puzzling. The manufacture of aluminum is a recent development. It could not be done until modern technology made it possible to generate the very high temperatures needed to purify bauxite. Yet, 3,000 years ago, people already knew how to manufacture it. How was this possible?

The ancient Chinese may have been practicing a form of alchemy, or in modern terms, atomic transmutation. Be careful not to confuse transmutation with transformation. Transformation is the process whereby an element changes from one form to another. For example, when subjected to high temperature and pressure, carbon powder will change into a diamond. Carbon powder has a completely different appearance than a diamond, yet both are comprised of the same element—carbon. In transmutation, an element such as carbon will change into a completely different element, such as iron.

According to modem understanding, atomic transmutation is more or less impossible. Physics and chemistry teach that:

1. Each element is separate.
2. Atomic transmutation occurs only under very special circumstances.

Scientists believe that transmutation may be occurring in the sun. They hypothesize that within the sun, hydrogen is changing into helium. However, the gases in the sun exist under extreme temperatures and pressures. Transmutation has been accomplished artificially in a cyclotron particle accelerator, in which high-speed nuclear particles are used to "bombard" stationary target atoms. This is the method by which heavy unstable elements, such as neptunium or plutonium, were made. However, conditions inside a particle accelerator are very extreme. High temperatures and tremendous centrifugal force must be generated. Moreover, these huge devices require a substantial investment of funds and manpower to build and maintain, and use violent, rather than peaceful and natural, methods to achieve transmutation.

Scientists believe that transmutation occurs only under circumstances such as these. The big bang hypothesis, for example, states that all the elements have existed in their present form since the origin of the universe. With the exception of radioactive decay, elements do not transmute themselves into other elements. However, is this view in accordance with the principles of the law of change? Was each element formed under a different set of circumstances? Is the origin of each different? Or is it possible that, under a variety of circumstances, one element changed—either gradually or rapidly—into all the different elements?

The unifying principle teaches that everything changes. Yin attracts yang and yang attracts yin. Yin repels yin and yang repels yang. At the extreme, yin changes into yang and yang changes into yin. If we agree with these basic laws, elements must be changing into each other under natural conditions. If we do not agree with these basic laws, then for us, such things are impossible. We view everything as static and separate, rather than seeing the unity inherent in all things.

6

Professor Kervran's Remarkable Discovery

When the future history of science is written, the name Louis Kervran could rank above that of Galileo, Newton, and Einstein. Kervran, a French biochemist who died about fifteen years ago, was working for the French government in a study that took him to the Sahara desert. He was investigating the physical condition of a group of workers. Exact records were kept of what the workers ate and what they eliminated. Kervran discovered a puzzling discrepancy. The workers were discharging more potassium than they took in.

Kervran began looking for the cause. He began by studying the ratio of sodium to potassium. The ratio between sodium and potassium is a reflection of the overall balance between yang and yin in our bodies and in the foods we eat. Kervran found that the workers were eating more sodium than they were eliminating. He also found they were eliminating a great deal more potassium than they were taking in. The conclusion he came to was that sodium was changing into potassium in their bodies. But, being a classically trained scientist, he also doubted his discovery, since it contradicts the basic tenets of modem chemistry.

Soon after Kervran published his findings in a scientific journal, a New York newspaper ran a small article about his work. The article came to the attention of George Ohsawa. Mr. Ohsawa immediately recognized the significance of Kervran's discovery. He understood that transmutation is nothing more than the law of change that humanity has known and practiced for thousands of years. Lao Tsu taught the same law, as did Confucius. When Ohsawa visited France soon afterward, he met Kervran, and encouraged him to write a book about his research. That book is *Biological Transmutations*. After the book was published, a sensation arose in scientific circles. And the issue has been argued from many different sides. Kervran authored two other books on the subject of transmutation, *Natural Transmutations* and *Transmutation of Plasma*.

A belief in the possibility of transmutation dates back to the origin of modern science. In 1799, a French chemist by the name of Vauquelin observed a large quantity of lime in the daily excretion of chickens. He fed a captive hen a diet of nothing but oats in order to find out where the lime was coming from. He measured the amount of lime in the oats, and then fed the oats to the hen. He then measured the amount of lime in the excretion and the eggs of the hen, and discovered that it had increased by a factor of twelve. He hypothesized that lime had been created but was unable to explain how or why.

An English researcher by the name of Prout defined the transmutation of elements in 1822. He studied incubating chicken eggs and noted that the increasing amounts of calcium carbonate inside the eggs had not originated in the eggshells. In 1833, Choubard studied germinating seeds of watercress. He observed minerals in the sprouting plants that had not been present in the seeds. Vogel, in 1844, also studied the seeds of watercress. He placed the seeds in a controlled air environment and nourished them with a solution that was free of sulfur. He discovered that the plants that grew had more sulfur in them than the seeds did. He hypothesized that the additional sulfur had to have come from an unknown source. Other nineteenth century researchers, including Lauwes, Gilbert, and Von Herzeele, also conducted studies on transmutation.

In 1960, Professor Baranger, an associate of Kervran's in France, published his findings on variations of phosphorus and calcium in germinating seeds. He concluded that a transmutation had occurred, but was unable to determine the reason. A review of the work of Kervran and others, entitled *Energy Development from Elemental Transmutations in Biological Systems,* published in 1978 by the U.S. Army Material Technology Laboratory, concluded that in biological bodies, transmutation was most likely taking place at the cellular level:

> The works of Kervran, Komaki [an associate of George Ohsawa's], and others were surveyed; and it was concluded that granted the existence of transmutations (Na to Mg, K to Ca, Mn to Fe), then a net surplus of energy was also produced. A proposed mechanism was described in which Mg adenosine triphosphate, located in the mitochondrion of the cell, played a double role as an energy producer. In addition to the widely accepted biochemical role of MgATP in which it produces energy as it disintegrates part by part, MgATP can also be considered to be a condensed cyclotron on a molecular scale. The MgATP when placed in layers one atop the other has all the attributes of a cyclotron in accordance with the requirements set forth by E.O. Lawrence, inventor of the cyclotron.

It was concluded that elemental transmutations were indeed occurring in life organisms and were probably accompanied by a net energy gain (italics added). The theory, then, is well established. But what needed to be done was to demonstrate transmutation in the laboratory. Once that was done, the whole orientation of modem science would change. The best way to prove that transmutation is possible would be to conduct experiments that show one element changing into another. Modem science would then have to accept the law of change and the unifying principle. Once the law of change is accepted, its application to technology, industry, medicine, economics, and politics would follow. Then the whole world would change.

Now let us see how sodium can be changed into potassium. Sodium has an atomic number of 11 and an atomic weight of 23. The atomic number of potassium is 19 and its weight is 39. If we subtract the number and weight of sodium from potassium, we get an atomic number of 8 and an atomic weight of 16. The element that corresponds to these numbers is oxygen.

7

CHANGING SODIUM INTO POTASSIUM

George Ohsawa wanted to prove transmutation in the laboratory, but was at a loss as to how he could do this. After trying unsuccessfully to design an experiment, he decided to go on a brown rice fast until he discovered a solution. One night he had a dream. From the sky, among the clouds, a big hand appeared. Bolts of lightning shot forth from the fingers. They descended to earth with great brightness. As each bolt struck the ground, it exploded, leaving newly created elements behind. When Ohsawa awoke from the dream, he knew he had found the answer: electricity would be needed to make transmutation happen.

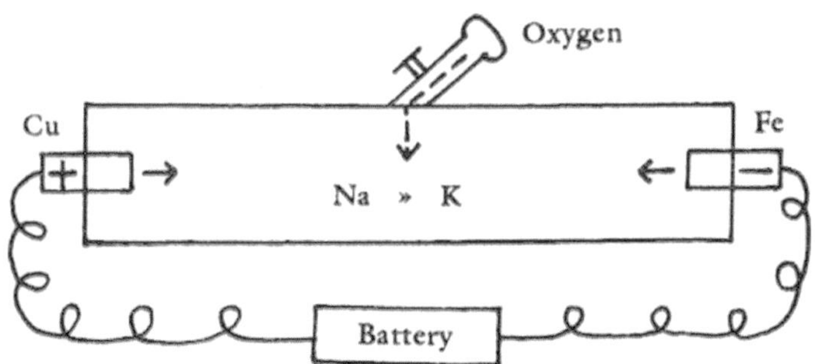

The Ohsawa Sodium into Potassium Experiment

Ohsawa contacted a friend who was a professor at a large university, and sought his assistance in setting up the experiment. The experiment was conducted in a very simple way. They took a vacuum tube and attached electrodes at either end; one positive, the other negative. They attached a wire to the electrodes. They put 2.3 mg of sodium inside the tube, at the middle of which they placed a bulb containing oxygen. They ran an electric charge through the wire, which, after some time, caused the sodium to melt into a liquid, gaseous, and finally a plasma state. They placed a prism in front of the tube, and an orange color appeared, a reflection of the more yang wavelength emitted by the sodium. At that point, they opened a valve, allowing the oxygen to enter the tube. Since sodium is yang and oxygen is yin, nuclear fusion between the two elements took place. They did not unite side by side as they do in a compound; instead their atoms fused and gave rise to new atoms. At that moment, the color coming through the prism changed from orange to purple, or from yang to yin. The result of the test was that 2.3 mg of sodium fused with 1.6 mg of oxygen and formed 3.9 mg of potassium.

This helped explain how the workers Kervran studied were able to change sodium into potassium within their bodies. The workers were taking sodium in the form of salt. Oxygen is abundant in the air. And if they were working hard and breathing heavily, there would be even more oxygen in their bodies. They were yang people doing yang work in a yang environment. Their bodies were highly charged with electromagnetic energy, similar to the electric charge passed through the vacuum tube. This combination of factors caused sodium to change into potassium in their bodies.

When I heard about Mr. Ohsawa's results, I began to see how the transmutation of elements directly reflects the spiral formation of the universe through seven stages, from infinity to the world of matter. In their experiments, Mr. Ohsawa's team had succeeded in duplicating the first five stages of the spiral of materialization. Each of the components in the experiment corresponds to one of these stages:

1. The vacuum represents infinity, or the infinite nothingness.
2. The plus and minus electrodes corresponds to yin and yang, or polarity.
3. The electric current corresponds to the world of vibration.

4. Plasma corresponds to world of preatomic particles.
5. The new element is a product of the world of atoms and elements.

Throughout the universe, these five worlds eventually give rise to the vegetable kingdom (6) and the world of animals and man (7). Now, if it is possible to change sodium into potassium, it should be possible to transmute other elements as well.

8

MAKING IRON

Iron is the most important element in modem industry. After the successful transmutation of sodium into potassium, the next step in our investigation was to discover a method for making iron. Let us first consider how the elements combine with each other and how they come into being through natural transmutation. We can start with the first eight elements in the periodic table.

Hydrogen can exist as a normal atom $^{1}_{1}H$ or as heavy hydrogen $^{2}_{1}H$ or $^{3}_{1}H$. Helium ($^{4}_{2}He$) is produced through fusion by the combination of $^{1}_{1}H$ and $^{3}_{1}H$, or by two atoms of $^{2}_{1}H$. Lithium ($^{7}_{3}Li$) arises from the combination of $^{4}_{2}H$ and $^{3}_{1}H$; beryllium ($^{9}_{4}Be$), from the combination of $^{7}_{3}Li$ and $^{2}_{1}H$; and boron ($^{11}_{5}B$), from the fusion of $^{4}_{2}He$ and $^{7}_{3}Li$. Carbon ($^{12}_{6}C$) arises from the combination of three atoms of helium $_{3}(^{4}_{2}H)$, or from other combinations of lighter elements. Nitrogen ($^{14}_{7}N$) is produced by the combination of $^{12}_{6}C$ and $^{2}_{1}H$; while oxygen ($^{16}_{8}0$) can be produced by the combination of $^{14}_{7}N$ and $^{2}_{1}H$, or by the fusion of $^{12}_{6}C$ and $^{4}_{2}He$. As we saw earlier, the magnetic elements—iron ($^{56}_{26}Fe$), cobalt ($^{59}_{27}Co$), and nickel ($^{58}_{28}Ni$)—occupy the fourth orbit of the spiral, which is the most balanced, central position. Since other elements are attracted toward this position, it should be simple to find combinations of elements to transmute into them.

The most basic element is hydrogen. Among the first eight elements on the periodic table, carbon is the most yang, oxygen the most yin. Oxygen has a very low melting point, much lower than atmospheric temperatures. That is why we know oxygen only in its gaseous form. Carbon, on the other hand, has a very high melting point. Most of the other elements have melting points in between these two extremes. Carbon and oxygen are the Shiva and Shakti of the elements; the primordial ancestors who can claim many of the heavier elements as their descendants. Our world is made up largely of three elements: carbon, hydrogen, and oxygen. Our food is also comprised of these three elements. We call this carbohydrate. If nitrogen is added to carbohydrates, protein is the result.

The fusion of carbon and oxygen produces silicon, according to the following formula:

Silicon-28
$$^{12}_{6}C + {}^{16}_{8}O \rightarrow {}^{28}_{14}Si$$

Silicon is a very hard, stable element, the result of the powerful bonding between two strongly polarized elements. The combination of two atoms of carbon and two of oxygen yields an element in the iron family:

Iron-56
$$_2(^{12}_{6}C + {}^{16}_{8}O) \rightarrow {}^{56}_{28}X$$

What is this mysterious element? It is a combination of iron, cobalt, and nickel. In order for iron to form, two protons must leave the new configuration, in a process known as double proton emission. As the transmutation is taking place, these protons disappear almost instantaneously, resulting in the formation of a stable atom, $^{56}_{26}Fe$.

9

THE CARBON INTO IRON EXPERIMENT

George Ohsawa used a graphite crucible for the carbon into iron experiment. Using carbon as the vessel for the experiment helped eliminate as many impurities as possible. Graphite is a form of carbon. Carbon powder was placed at the bottom of the crucible. A wire was attached to the carbon rod and connected to a battery. Another wire led from the opposite pole of the battery to the carbon crucible. When the carbon rod touched the carbon powder, sparks were produced, causing each fleck of the powder to reach a highly charged, plasma state. The necessary oxygen entered from the air, and automatically combined with the carbon. Nuclear fusion arose, and the result was the production of nickel, cobalt, and iron. Nitrogen, carbon dioxide, and other elements are also present in the air, and if these interact with the carbon, other elements will also be created.

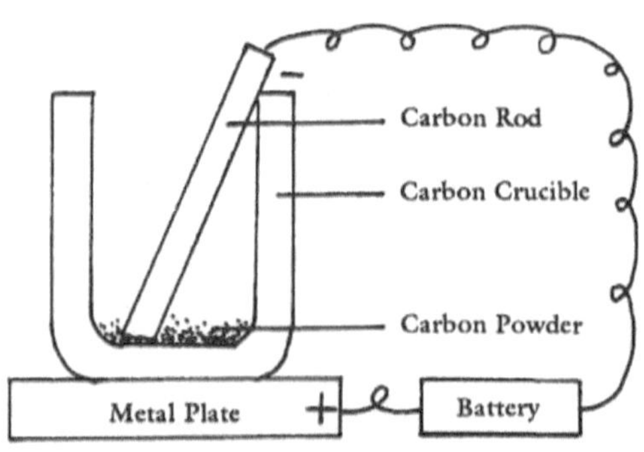

The original Carbon Arc Experiment

After doing the experiment a number of times, I noticed that when I let the time for the transmutation get longer, more iron resulted. The same result will happen if you control the temperature. If the room temperature is colder, more iron will result. Yin, or cold, accelerates the coming together, or fusion, of atoms, a process we can refer to as *cool fusion*. The iron produced through transmutation has a very special quality. Because it is formed from the combination of yang and yin, or carbon and oxygen, it has a very stable quality. This stable quality makes it resistant to rusting. It is also very hard, and is resistant to heat. Iron with these properties can be very important in industry, for use in railroads, shipbuilding, and construction.

This is transmutation at low energy, temperature, and pressure. Contrasted with present scientific understanding of nuclear physics, it seems miraculous, but it is only a simple technological application of yin and yang. In nature, the attraction of yin and yang is greatest between oxygen and carbon. It is this attraction that gives rise to iron. The electrical energy released in sparks activates the carbon atoms on the dish and the random oxygen molecules that are in the surrounding air. As they begin to cool rapidly and contract, they fuse and form iron. Yang energy produces yin—expansion and excitation; then yin, cool temperature, causes yang—condensation and fusion.

Since such fusion reactions can occur at low energies and temperatures, they are common in nature, particularly in plants and animals. In fact, without the theory of low-energy transmutation, it is impossible to explain the origin of the first cellular life. According to modem theory, the division of prior cells produces cells. Where did the first cells come from? Some speculate that bacterial or viral spores may have been transported between solar systems. However, this avoids the basic question: how did living matter organize itself from the inorganic world of elements?

To understand the law of transmutation, we are studying yin and yang. Within this study we can grasp the underlying unity of life overlooked by modem science. Thus we can not only see the relative and frozen world of matter, but also glimpse its mutability.

10

RARE METALS

The three most important transmutations are: 1) from sodium to potassium, 2) from carbon to iron, and 3) making precious metals, such as gold and silver, from more commonly available elements. If you can do these transmutations, even in small quantities in your home, the impact on the modem world will be great. Because if you are able to produce these elements in small quantities with simple equipment, then it is possible to make them on an industrial scale. Once we have learned how to make metals, and then how to use them, the whole of civilization will change. Material civilization will change. Now that materials must be dug out of the ground and are hard to obtain, we place special value on them. We are materialistic. We are attached to these metals. But if it becomes possible for everyone to make metals in their home, if sheets of metal become ten cents apiece, then materialism will vanish.

Some people believe that freedom comes from eliminating desire, eliminating the goods which they may want, or from going into the mountains and meditating. That way is very beautiful. I respect that way very much, but that method would make me feel very lonely. I would rather follow a way that would make these things so abundantly available there would be no more attachment to them. Whatever you want, you can get. Whatever you don't want, you don't use. That is actual freedom from the material world. I think you have heard that in the Medieval Age alchemists made gold. And you have often wondered how they did this, or if, in fact, this was but a fictional story. In the Orient, there also exist stories that the graduates of Lao Tsu's school also made gold.

In both Eastern and Western accounts, the metal that was used in the beginning stages of the gold-making process was mercury. Gold is brittle and yellow. Mercury is silver and liquid. They are very different. But amazingly enough, the alchemists knew that if they applied the proper technology, the could transmute mercury into gold. When you locate these two elements on the periodic table, you will notice that gold has the atomic number 79 and mercury, the atomic number 80. They are next to each other. In order to effect this change, Western alchemists used what was called the philosopher's stone. So far in the modem age, no one has discovered what the philosopher's stone was made of. We can guess at its composition, but we cannot know for sure what it was, as it has disappeared.

However, when Mr. Ohsawa was doing the experiment transmuting sodium into potassium, he tried different temperatures and different lengths of time in order to see how that would affect the results. He began to notice some strange metallic impurity in the final potassium. He was curious to know what this impurity was, and had it analyzed at the laboratory. Strangely enough, it was gold—a very small volume, indeed, but nonetheless, gold.

This result was puzzling. How did the gold appear? The atomic weight of gold is 197. To figure out how this happened, add the atomic weights of all the elements used in the experiment, including the iron and copper in the electrodes, the sodium and potassium in the experiment, and the oxygen in the air. The result is 197—the atomic weight of gold. Nuclear fusion of all of the elements in the vacuum tube had produced a tiny amount of gold. So then, we know that gold can be made by transmutation. Even though this result was not produced by the legendary practice of starting with mercury, still gold was produced.

Elements Used in the Sodium to Potassium Experiment

O_2	16
Cu (in electrodes)	63
Fe (in electrodes)	56
Na	23
K	<u>39</u>
Total	197

I was still curious how the ancient alchemist-philosophers transmuted mercury into gold. I decided to experiment with an open crucible, because the ancients did not have vacuum tubes. They probably used some type of iron containers. I bought several small vials of mercury. Mercury has an atomic weight of 200, gold, an atomic weight of 197. So far in the transmutation experiments, we combined or added the atomic weights of two or more elements to produce another. In this case, we have to subtract. This presented a new difficulty. I did not want to use a destructive method to achieve transmutation. I did not want to use a huge particle accelerator, with enormous centrifugal force, generating very high temperature. Rather I wanted to use the methods of harmony. I preferred to understand the natural attractions between elements and employ them, so there would be no excess of energy or radioactivity produced.

After studying this problem, I began to see several possible means for transmuting gold using the addition method. But unless I could understand the subtraction method of transmuting mercury into gold, which is what the ancient alchemists used, then my understanding would have remained incomplete. If they could do it, then using our understanding of yin and yang, we should be able to do it also.

After struggling with this problem for some time, I came to certain conclusions and began to experiment with them. I failed several times. The big question in my mind was still the philosopher's stone. I began to understand that it was not a natural stone. Rather it was a man-made stone. I deduced, if my interpretation is correct, that the philosopher's stone must be made from phosphorus. Phosphorus is yin. But for this process, the phosphorus must not be yin; it must be made yang. Heat must be applied to the phosphorus to change it into a deep ruby color, and it becomes very hard like a stone.

I had no big laboratory, so I had to do my experiments in the kitchen, using household current of electricity, using kitchen pots and pans, and the gas stove. Of course, this is as it should be. That is the equipment and facilities used by ancient alchemists for their experiments. We have one advantage, however, over the ancients. We have an ample supply of household electricity that can be used to speed up the heating factor in the process. We know from legend that it took quite a long time for them to complete the transmutation of gold. With the aid of electricity, the time can be shortened.

As a result of these early experiments, I feel I have graduated from the school of Lao Tsu, the goal of which was the application of the law of change to the practice of alchemy. But, I am not yet graduated as a modem alchemist. I am graduated as an alchemist, an individual alchemist, but in order to be a modem alchemist, I must discover a method of mass production. That is my future homework.

11

VALUE

In the 1956 spy novel, *Diamonds Are Forever,* author Ian Fleming describes the mythical allure of diamonds, in a scene in which British secret agent James Bond examines a twenty-carat blue-white diamond:

> Bond put down the piece of quartz and gazed again into the heart of the diamond. Now he could understand the passion that diamonds had inspired down through the centuries, the almost sexual love they aroused among those who handled them and cut them and traded in them. It was domination by a beauty so pure that it held a kind of truth, a divine authority before which all other material things turned, like the bit of quartz, into clay. In these few minutes, Bond understood the myth of diamonds, and he knew that he would never forget what he had suddenly seen inside the heart of this stone. He put the diamond down on its slip of paper and dropped the jeweler's glass into the palm of his hand. He looked across at M's watchful eyes. "Yes," he said. "I see."

Man is a very interesting being. He will put a great deal of value in a useless object, such as a glittering diamond or a bar of gold, or in ornaments that are worn as jewelry. These things have little or no actual use. Industrial diamonds are used to make holes or cut things, these are more useful. But these very expensive, "valuable" things are used just for ornamentation. In comparison with this, one pound of brown rice is far more valuable for life. You can eat brown rice. But you can't eat gold or diamonds. Yet people will pay hundreds of thousands of dollars for a single diamond. And brown rice is very cheap. It is very interesting. Values are so conceptual.

If you discover a simple way to make diamonds or gold, then for you the material world is no longer meaningful. These once held material values would be changed. What will remain then is the ability to use the material world freely.

12

PLATINUM

Which is higher priced, platinum or gold? Platinum. What is the atomic number of platinum? Again it is next to gold. The order is platinum, gold, and then mercury. The atomic number is 78. And the atomic weight is 195. Platinum is an expensive metal. If we wanted to make platinum, how would we do it? What elements would be required? Again several different methods are possible. But the simplest one uses only two elements: carbon and oxygen. As we have seen, these two elements combine very easily and readily. The combination of one carbon atom with one oxygen atom will result in one atom of silicon. The combination of two atoms of carbon and two atoms of oxygen will result in a mixture of iron, cobalt, and nickel. The combination of three carbon atoms with three oxygen atoms will result in krypton. When we finally combine seven atoms of carbon with seven atoms of oxygen, the result we obtain is an isotope of platinum and gold. The product is somewhere in between platinum and gold. But it has a strong tendency to become platinum. So you can see that platinum can be made very easily.

Carbon and oxygen are readily available on the earth. Among the first eight elements, they have the greatest degree of attraction. In the evolution of the elements, it is the interaction of these two that form the other elements. Carbon can also be replaced by boron. In terms of yin and yang, these are like brothers in that they react in similar ways in forming new elements. The natural evolution of the elements follows the order described above, the combination of yin and yang continues for seven octaves or movements:

1 (C + O) $\rightarrow$ Si (atomic weight 28)
2 (C + O) $\rightarrow$ Fe (56)
3 (C + O) $\rightarrow$ Kr (84)
4 (C + O) $\rightarrow$ Cd (112)
5 (C + O) $\rightarrow$ Ce (140)
6 (C + O) $\rightarrow$ Er (168)
7 (C + O) $\rightarrow$ Pt (196)

In the Orient, the number seven is very active, or yang; the last number before the return to infinity, which is symbolized by the number eight. In the same way, the seventh octave is the end of the spiral of the elements. Platinum stands as the last transmutation of carbon and oxygen. After that, the elements become radioactive and start to decay. Beyond platinum, gold, and mercury, more and more of the elements are radioactive and unstable.

We might call radioactive decay, death. The "matter" of the element returns to a simpler state as a lighter element. As this happens, the element's yin components are discharged as invisible preatomic particles and energy. But death is a part of the cycle of life, for particles will reassemble and transmute the heavy elements again. Is this not like biological life?

The grand cycle of birth and death of the elements we call the symphony of carbon and oxygen. It begins with a prelude: the first eight elements that suggest the major notes of the symphony; then there are seven movements during which are built variations on the original theme. More and more intricate the melodies become as the elements grow larger and heavier. The "surroundings" are the conductor whose intensity is deepened through seven movements until finally there is a long coda, as the elements dissolve into the silence of infinity. Yang changes into yin. As the symphony of the world of matter arrives at its conclusion, those who listen may hear the eternal rhythm of infinite motion, which is called the order of the universe.

13

PURE CARBON

While I was doing the experiments with carbon transmutation, I wanted to have the carbon rod as pure as possible. If impurities were there, it would have been difficult to track the results of the experiment. I checked with several chemical companies who were producing carbon powder and carbon rods, and found that 100 percent carbon rods and 100 percent carbon powder do not exist. Impurities are always there.

Many elements can be taken out in order to make pure carbon. The boiling temperature of carbon is extremely high, so when heat is applied to carbon, the other elements boil away. In this way we can make nearly 100 percent pure carbon. And yet we cannot make completely pure carbon, no matter how hard we try. Especially one element is nearly impossible to eliminate from the carbon. It is always found in carbon rods and carbon powder, even though it may be a tiny amount. That element is silicon. It is very interesting how this silicon comes to exist in the carbon rods and powder.

Please check the atomic number and weight of silicon (Si). Why is it practically impossible to separate the silicon from the carbon? Carbon is a very yang element. Oxygen is all around the carbon in the air. Carbon and oxygen have the tendency to attract each other. Usually they are forming compounds such as carbon monoxide and carbon dioxide. But at the same time, nuclear fusion happens, and with the combination of one carbon atom with one oxygen atom, an atom of silicon is formed. So as soon as we make purified carbon, at the next moment silicon is already being formed. This phenomenon occurs with many metals—they can make them nearly pure, but traces of other metals will always appear.

14

RADIOACTIVE ELEMENTS

In order to understand more about radioactivity, we need to consider the way in which elements change and develop within the solar system. The solar system is made up of nine planets. There may be other planets which are still undetected which may eventually come into the solar system. If we call the distance from the earth to the sun 1 astronomical unit (1 A.U.), then the distance from the sun to Pluto is about 40 astronomical units (40 A.U.). Beyond the nine planets lies a vast field of comets. According to one estimate, there are about a hundred million comets orbiting at the periphery of the solar system, making the sun as the center of their orbit. This cloud is about 3,850 times larger than the field of planets.

If we see the complete solar system, the sun is in the center, the nine planets revolve around it, and beyond this central nucleus spins the cloud consisting of the comets. If we include this vast cloud, the solar system on the whole is about four light years across. The entire configuration takes the form of a logarithmic spiral with seven orbits, a form we see duplicated in shells such as the chambered nautilus. The forces of centripetality and centrifugality, or yang and yin maintain this spiral. As it does in the atom, the more yang force of centrifugality pushes the particles at the periphery, in this case the comets, in toward the center. Comets eventually spiral inward and become planets, while planets spiral inward toward the sun.

The present distance between the earth and the sun is about 93 million miles. Sometimes the distance is a little greater and sometimes it is a little less. But the average distance is 93 million miles. On the whole, however, this distance is actually becoming a little less, since the earth is slowly approaching the sun. All of the planets are traveling toward the sun. That is how the sun is fed. The sun is burning. The sun is not a planet. Rather it is the huge melting spot at the center of the solar system.

According to modem thinking, the sun is a body of burning gas. It is, on the contrary, not a "body," but the melting point of the solar system. It is the center of the solar system where all things disintegrate. If we compare the structure of the solar system to that of the atom, the comets correspond to electrons, the planets correspond to neutrons, and the sun corresponds to the proton.

A very long time ago the earth was in the same position as Jupiter is now. In between these two positions is the asteroid belt. The asteroid belt is made up of many small bodies. Jupiter has thirteen satellites, Saturn has ten. Mars has two small satellites, while the earth has one. Venus and Mercury have no satellites. Why does such a big gap arise between Jupiter and Mars?

When we bum a candle, the flame rises in three different colors. We can compare the flame of the candle to the arrangement of planets around the sun. The innermost flame represents the first four planets: Mercury, Venus, Earth, and Mars. The second flame represents the next two planets: Jupiter and Saturn. The third flame represents the last three planets: Uranus, Neptune, and Pluto. Each group of planets is at a different stage of development. When the planets pass from the middle to the inner stages, they pass through the asteroid belt.

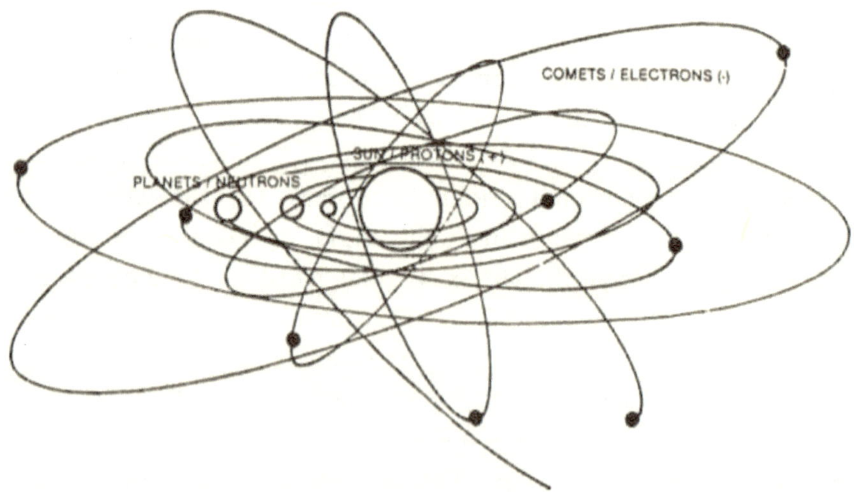

Atomic model of the solar system

The asteroid belt possesses a strong electromagnetic charge, and the many small bodies within it are moving very fast as a result of this strong charge of energy. When Jupiter passes through the asteroid belt, it will become much smaller, and will lose many of its satellites. Jupiter will, in other words, become more yang. In this process, peripheral satellites, which are yin, are discharged one by one, until no more satellites will be there.

Before the planets passed through the asteroid belt, there was no life on them. Once the planets have passed through the belt and after perhaps some millions of years, when they have reached the position of Mars, then life begins to appear. Because of the increase of solar radiation and because of the electromagnetic charge received during passage through the asteroid belt, life begins. Mars is colder than the earth. Yet, primordial life—in the form of microscopic bacteria—may be present there, although it has not yet been detected. As Mars approaches the position of the earth, then more and more life will flourish on it. The time it took for the earth to progress from the position where Mars is now to where the earth is now took about 3.4 to 3.6 billion years. And organisms, not just microbes, but living organisms, began to appear on the earth's surface about 3.2 billion years ago.

The sun is composed mainly of hydrogen and helium. On earth there are 107 elements. After, generally, the 90th element, the remaining ones are radioactive. Hydrogen and helium are atomic numbers 1 and 2. The earth's elements, as we have studied, have been created from the hydrogen atom. Now, as the earth approaches the sun, perhaps after several billion years, the elements on earth will begin to decompose. Additional elements will start to become radioactive. As the earth comes close to the sun, then not only will elements from 90 on be radioactive, but the elements from 80 on, and then from 70 on, from 60 on, and so forth. Gradually, more and more elements will become radioactive, including more balanced elements such as iron, cobalt, and nickel.

Finally, as all the elements approach the point of radioactivity, the earth will assume a cloud-like form and be composed entirely of light elements such as hydrogen, helium, oxygen, and nitrogen. The earth will eventually enter the sun, at which time a hydrogen explosion will occur. That is sunshine, solar radiation.

You can imagine that in the outer solar system, especially in the very distant planets—Uranus, Neptune, and Pluto— elements that are currently radioactive on earth are at this point stable. These outer planets may even have more elements than we know on earth today. They may have as many as 120 or 130 elements. Uranium is currently radioactive on earth. But on Mars, uranium may not be radioactive. Following the same progression, carbon may be radioactive on Venus or Mercury.

Like everything else, the elements and planets are subject to the law of relativity. Elements may not be the same on other planets as they are on earth. The law of the order of the universe is the same throughout the entire universe. But actual phenomena are always different. So it is not important to memorize a great deal of data. Rather we should try to understand the universal law that gives rise to all phenomena. Among all forms of knowledge, the understanding of the law of change is the most important.

15

APPROPRIATE TECHNOLOGY

The exercises presented below are a continuation of the chemistry studies in the earlier part of the book. They represent a more technical, practical application of yin and yang. Whether we realize it or not, in dealing with technology, we are always applying the principles of yin and yang. Without the application of yin and yang, no technology can work. However, once we understand yin and yang we can solve technological problems very efficiently and economically.

Many of us do not understand the intricate mechanism of modem technologies. Many of us would prefer to abandon modem technology and become closer to that which we feel is more natural. And many of us are inclined toward artistic, aesthetic, or philosophical fields rather than engineering or other technical fields. But once we understand that all phenomena in the universe are governed by yin and yang, and once we understand that all the things we are doing (like eating, walking, sitting, and cooking) are all subject to the principles of technology, then we will be better able to confront each problem with a clearer approach, a more simple understanding. Let us now apply our understanding of yin and yang to the technical problems presented below.

1. Suppose you have an egg. The white is yin, the yolk, being in the center and yellow in color, is yang. Now, you would like to make the yolk hard, but you would like to leave the white part soft. What should you do? According to yin and yang there is a very simple

solution. You should make cold. Do you understand why? The yolk, being yang, will attract the cold and freeze hard. But the white, being yin, will reject the cold and remain soft.

But if you want to have the opposite reaction: to have the yolk soft and the white hard, then you apply the opposite factor—heat. You cook it. Of course if you cook the egg for a long time then everything will become hard. But with a small amount of heat the white will become hard while the yolk remains soft.

2. Any country will serve as an example for this problem, but we will look at Russia. Close to the Black Sea it is warm and many fruits grow there, but near Moscow it is very cold, the winter is long and not conducive to the cultivation of fruits. In order to grow fruit trees about nine months of warm weather is required. It takes time for the juicy fruit to grow and ripen, and then, in late summer or early autumn, it is ready to eat. But we would like to try anyway, even though we know that fruit grown in Moscow will probably be smaller and less juicy than fruit grown further south. We would like to take fruit trees from the warmer, southern climate and transplant them near Moscow. In order to do that, the fruit must ripen during the brief warm period. So the trees will have to grow rapidly. Suppose we have 100 seeds from the Black Sea region. What should we do with them?

First we have to select the most yang seeds, the smallest. Then we must put them into cold storage. This will make the seeds more yang and tight. Then in Moscow when the snow melts in May we can plant the seeds. These cold contracted seeds, which are compact and thirsty, will readily take in the water of the melting snow, and sprout quickly and grow. If we do not put the seeds into cold storage, they will remain very yin. They will not be thirsty. They will grow very slowly, not being able to bear fruits before the autumn frost. So the important thing to remember is to make the beginning very yang.

3. Once there was a Chinese farmer who understood yin and yang. He raised pigs on his farm. His farm was very small. A more prosperous

farmer lived next to him and had a bigger farm and raised large, award-winning pigs. The small farmer decided to apply his understanding to produce pigs that would be even larger than those of his neighbor. What did he do?

When the pigs were very young, he fed them very yang food: buckwheat and burdock, and not much water. After they were several months old, he widened their diet and gave them more water. Then the salt and minerals they retained from the yang food attracted plenty of liquid. They grew rapidly and became quite fat.

4. A long time ago, a young Scotsman by the name of James Watt saw the cover of a steam kettle move when the water in the kettle was boiling. He started to think that maybe this power could be used to make a piston move. The steam from the boiling water could be used to drive a wheel, converting this water to a usable source of energy. This was the creation of the steam engine. Watt did not know yin and yang; he just observed the phenomenon and developed a technology based on it.

But suppose you know yin and yang. If you saw this boiling kettle, what would you think? You must have a better idea than James Watt. You should employ the water at the bottom of the pot. That is the most yang water. Fill a test tube half full with water. Cap it tightly with a stopper or cork. Heat the water in the tube until it boils. When the steam becomes very active then the stopper will pop out. That is the principle with which James Watt worked. But before the water boils, after the water has become somewhat heated, you can slowly tip the test tube. When the hot water reaches the stopper, it will pop out. This is a more yang method of obtaining power. The power contained in heated water is very large. Heated water is more yang than steam. It contains far more power.

5. In order to feel cool in the summer, what color clothes should we wear? We should wear white. In order to stay warm in the winter, what colors should we wear? We should wear black or dark. The energy of the sun is very yang. White is a yang color. Yang repels yang. Therefore, the sun's energy—heat—is repelled. The energy

of winter or snow is yin. Black or dark colors are also yin. Yin repels yin, so the cold is blocked and we feel warm. If you wear black in the summer, then you will retain heat, since yin attracts yang. The same thing is true during the winter. If you wear light colors in winter, you will feel colder. Yang attracts yin. Even without knowing it, we often select colors according to the principle of yin and yang.

6. Suppose a group of doctors want to discover a better method for giving injections, so that the body will more efficiently utilize the medication or nourishment. When a nurse gives an injection, she rubs the skin with alcohol in order to make the area clean. But alcohol makes the skin contract. It is more difficult for the needle to penetrate the skin. Injections are mostly given at the periphery, or the most yin part of the body. At the periphery there is a tendency for energy to flow outward. This makes it more difficult to put something into the body. The mouth is more yang, and therefore a more appropriate place to take nourishment into our bodies. The periphery is a more appropriate place for discharging excess. Therefore, the treatment of disease by means of eating proper food is technologically sounder than giving injections.

7. Current thinking about the problem of how to convert solar energy into a source of household heat utilizes an enclosed water system that requires a storage tank, a radiator in each room, a connecting pipe to carry the water out to the solar energy collector (behind the collector, the pipes may be coiled for maximum space coverage), and a connecting pipe back to the storage tank. Is there any way to improve this technology?

    The problem with solar energy collection is that it works only during the daytime. So at night there is a lack of heat, and to correct this they are thinking about supplementing the heat with wind generators. But there may be no wind to drive the wind generator. For a source of energy to be effective, it must be constantly there. Within all types of energy on the earth are just two basic currents: the centripetal energy showering down upon the planet, generated by the infinite cosmos, and centrifugal energy produced by the earth's rotation. We refer to these two currents as heaven's force and earth's force. These two currents create all other types of energy, such as sunlight, wind, fire, hydropower, electricity, natural gas, petroleum, coal, and atomic energy. Every possible form of fuel or energy we could use on earth is nothing but a byproduct of these two original forces. "Clean" or "efficient" energy means that heaven/s and earth's forces are used as directly as possible, with their indirect byproducts used as little as possible.

These primary forces themselves—heaven's more yang, centripetal force and earth's more yin, centrifugal force—are actually the most ideal, natural source of clean energy. But it is very difficult to figure out a means to tap into this source of energy and harness the forces of heaven and earth. And yet our bodies catch this energy all the time. We are constantly being regenerated by these primary energies. That energy is constantly charging us, while we are sleeping, while we are eating, no matter what we are doing. If that charge stops then we have to die.

Our cells and organs are constantly being charged by this energy. It is constantly there. In comparison to solar energy, heaven and earth's forces are far greater. The primary force of yin and yang causes the earth to rotate, holds the moon in orbit, and drives the motion of the sun and planets. But the technology is difficult. I hope that we can someday invent a suitable technology. Below are two ideas for pursuing the goal of unlimited natural energy.

While the earth rotates, throwing off energy from the equator, tremendous streams of celestial energy are also coming in through the poles. These streams are coursing through the entire planet, forming huge lines of electromagnetic energy; along these highly charged energy belts, mountain ranges have been formed, running from north to south. Some mountain ranges run east to west. These were formed prior to recent shifts of the earth's axis and show the outline of the earth's previous north-south orientation.

As we know from acupuncture and Oriental medicine, the

energy in the body's meridians surfaces at various points, which are actually like tiny holes. These points are used to adjust body energy, either pouring in more energy if a deficiency exists or draining off energy if the point is excessively charged. These points correspond to the earth's volcanoes, such as Mt. St. Helens and Mount Vesuvius. If we can develop a method to channel into these points, as we do in acupuncture with a fine needle, energy could be drawn off and distributed to the entire world. As long as the earth is rotating, and the cosmos exists, this energy source is unlimited.

Moreover, within the continuous stream of energy radiating from the sun to the earth, are many preatomic particles, including ionized protons and electrons. As this current, known as the solar wind, approaches the earth, it separates: the more yang particles, such as the positively charged protons, are attracted to the more yin evening side of the earth, while the more yin, negatively charged electrons are attracted to the more yang morning side of the earth. At the side of the earth facing directly away from the sun, high above the atmosphere, these polarized streams of plus- and minus-charged energetic particles collide with tremendous force. As long as the sun is radiating toward the earth, this powerful collision and energy-generation takes place. If we could place several satellites into orbit, so that at any given time one would be passing through this highly energized zone, we could send down to earth a continuous stream of vast, unlimited energy.

8. Within the world of plants, certain varieties are yin, others, yang. Bamboo and pines have complementary-opposite structures. Why is the trunk of a pine tree solid, while the inside of bamboo is hollow? Bamboo grows in warm climates. It grows rapidly, which is more yin. In the beginning it will grow one or two feet each night. The spiral formation of the trunk has no time to develop into a solid mass, but remains only on the periphery as the spiral moves rapidly up the new growth. On the other hand, the pine tree, which grows in a more yang manner in a cold climate, has the time to allow the spiral to compact very tightly inside the trunk.

A similar difference can be found in the organs of the body. Organs like the bladder and intestines are yin or hollow, while the kidneys, spleen, and liver are yang or compact. Which type of organ develops more rapidly? The hollow ones develop rapidly. It takes longer for the yang, compact organs to develop. The kidneys, which are very compact, take almost the entire embryonic period to form, while the hollow digestive vessels develop very rapidly. Bamboo leaves are flat and well developed, while pine needles are thin, sharp, and contracted. Why do bamboo leaves and pine needles have these particular structures?

Temperature is the major cause of this difference. In order to make a liquid become solid we have to apply cold. Heat makes water and other liquids evaporate. Bamboo leaves are an expanded version of pine needles, while pine needles are a contracted version of bamboo leaves. Pine needles correspond to the veins in the bamboo leaf. If you can imagine the fleshy part surrounding the pine needles, you can see how that can be. Sharply pointed leaves, such as those of elm, maple, and oak, have a form that lies in between these two extremes.

Now, using yin and yang, let us figure out what kind of root structure these two have. Making the ground as a dividing line, the roots go down. The roots of the pine tree branch out, but remain contained within the specific area of the pine tree. Bamboo grows straight up, therefore the roots grow in the opposite direction; they spread horizontally below the ground.

Now let us suppose that there is a grove of pine trees and a grove of bamboo. Suddenly an earthquake begins. Into which grove should you run? You should head for the bamboo grove. Since the bamboo roots spread horizontally under the ground, it is more difficult for the earth to separate. You will be safer in the bamboo grove. Pine trees, on the other hand, will quickly uproot and could easily topple over. However, suppose a forest fire has begun. Into which grove should you run this time? Since the bamboo is hollow, it will bum quickly. Pine, on the other hand, is more compact and will not bum as easily. You should go to the pine grove. Again, you must use yin and yang.

9. This is a more difficult problem. When you visit Oriental countries, such as Thailand, Burma, and Japan, you will see many pagodas. In China and Southeast Asia there are not many earthquakes, so pagodas can stand more easily. But in Japan, earthquakes are a daily occurrence. So pagodas require particular architectural attention. Over a thousand years ago, a wise carpenter built a beautiful large pagoda in the city of Nara, an ancient capital of Japan. It is still standing today, even though there have been many earthquakes in that region. Many other buildings have fallen down, but the pagoda is still standing. What did the wise carpenter do to support the pagoda?

   The carpenter suspended a wooden pole from the top of the pagoda; it is free to move about, and there is a space separating it from the floor. It can swing about like a pendulum. When the pagoda moves with the force of an earthquake, the pole moves in a circle, making a center of gravity. After the earthquake is finished, then the pole will return to the center. In addition, the pagoda rests upon soft earth, which does not have the tendency to break up during a violent upheaval.

10. When we heat water, the volume becomes enlarged, bigger, changing to evaporation at 100 degrees C. When you lower the temperature, it contracts until it reaches 4 degrees C, after which it expands. At 0 degrees C it turns to ice and will break a glass container. Why does water increase its volume in both cases—whether we apply heat or cold? Daily you observe this happen. This is a scientific riddle.

    The hydrogen atoms in a water molecule are smaller than the oxygen atoms. They have a similar quality so they do not combine. Hydrogen is more yang; oxygen is more yin, so it is on top of the molecule. One hydrogen has a plus charge, the other, a minus charge. When we heat water, the oxygen more rapidly becomes yang, and begins to repel the more yang hydrogen, resulting in separation. Conversely, when we apply cold, the hydrogen quickly becomes yin and is less attracted to the oxygen, again, forming expansion.

16

Guidelines for Appropriate Technology

Like everything else, modem technology has a front and a back, a positive and a negative side. We all appreciate the technical conveniences of modem life, yet because of these conveniences, humanity is facing an environmental crisis. Global warming, depletion of the ozone layer, and accumulation of toxic materials in the soil, water, and air are only a few of the environmental problems that confront us at the beginning of a new century.

Modem technology is wasteful and inefficient. It depletes our non-renewable resources and pollutes the environment. In developing the technology of the future, we can use the understanding of yin and yang to avoid these pitfalls. The following points can help guide us as we develop self-sustaining technologies:

1. Technology must not weaken humanity.
2. Over-technology is the same as insufficient technology. In the same way, being too clever is like being too foolish.
3. The most important technological development is strong people, a strong humanity. As an example, when people feel cold, we put in central heating—70 degrees every day. Then people become weak. Instead, technology should make people stronger—so they do not feel the coldness much. What do we call this technology? For example, there are two solutions to the energy crisis. One is to

use the earth's resources for energy. The other is to make people content with limited energy. What do you call this second way? Macrobiotics, what we are doing. Making a strong autonomic nervous system—parasympathetic and sympathetic—and making judgment high and developing physical endurance; these are all part of macrobiotics. There is no other such technology in the world.

4. Technology should not waste principal. The earth's non-renewable resources are our principal. Every year trees are growing, rain is falling, and winds are blowing. Those things are interest. If we use only the interest, that is fine. But if we use the principal, the principal becomes smaller, the interest becomes smaller, and eventually we become bankrupt. Modem humanity has abused this principal, by digging out coal, oil, minerals, and other natural resources. Until the middle of the nineteenth century, humanity did not touch this principal. For example, charcoal, which is made of wood that can be replenished, was commonly used as fuel. Technology should use either interest alone, which was used for millions of years, or it should use unlimited sources. What are unlimited sources? Gas and oil are limited; eventually we will use them up. Unlimited sources include wind, sun, electromagnetic energy, the oceans, geothermal heat, and sound.

Suppose we tap into unlimited sources of energy. Then, what kinds of raw materials will we use this energy to build with? For the most part, civilization has been using nonrenewable resources— the principal. In addition to unlimited energy, we need an unlimited source of raw materials. Deposited sources are limited. How can we secure unlimited sources of raw materials? That is where transmutation comes in: in other words, we are able to make iron, gold, copper, and other rare elements from air, water, and other substances that are commonly available. Then, without having to touch the capital, civilization can develop without limit.

Now, let us consider a new problem. Suppose we secure unlimited energy and materials. Then for what purpose will humanity use this? Perhaps for guns or atomic bombs. That brings us to the next principle.

5. Technology should not be the leader. Technology should follow. What, then, is to lead? The order of the universe, or in other words, justice and love, should lead humanity. Without spiritual, philosophical, ideological under standing, an unlimited source of energy and materials could easily become a monster that destroys humanity. In the future, we must solve these problems. Our proposals: first, let us lea n how to secure unlimited energy. Second, let us learn how to secure unlimited materials. Third, let us learn to control energy and materials. In other words, we must develop our understanding of justice and love and what constitutes human happiness.

   If we have philosophical understanding alone, fine, but without technology the earth will be a primitive place. And energy and materials alone can destroy humanity. Spirituality and technology must go together, and spirituality must take leadership; technology must follow. In modem civilization, technology is taking the leadership, and spirituality follows. As a result, humanity is declining and our environment is being spoiled.

6. Technology has front and back. For instance, if a bomb should fall on Niagara Falls, then the lights would go out all over New England. The Arabic nations stop the oil supply, and there is a worldwide shortage of energy. Very inconvenient, isn't it? Something happens in one place; then the whole world is paralyzed. The more wonderful and universally affecting technology is, the greater back it has. Once it goes wrong, it affects or destroys everyone.

   The more technology develops, the greater the danger becomes. X-rays are very convenient, but because of X-rays, people are affected by radiation. Insulin is very convenient for diabetic people, but once they begin to take insulin, it becomes very difficult to cure the disease—for their whole life they must take insulin, in most cases. Very wonderful and convenient technology has a great back. We must understand the front and back, or positive and negative aspects of technology very clearly.

17

PATTERN FOR THE USE OF ENERGY

There is a definite pattern for the use of energy, a pattern, which modem civilization is not following. When this pattern is established, then civilization is accomplished. The pattern for using energy corresponds to the ratio of foods we take in. The food we need in the smallest quantity is salt, minerals. Then, in a quantity seven times greater than minerals, we take protein. Next, seven times more than protein, we take carbohydrates. Next, seven times more, water.

Seven times more than water, we take air. Seven times more than air, we take various vibrations. Seven times more than vibration, we take electromagnetic current, or infinite force. There are seven stages of food that we take in a one to seven ratio. Food is the source of human energy. For technology, for the building of civilization, the same ratio must be followed.

1. Salt and minerals correspond to deposited resources, such as coal and oil. These are to be used in the smallest amount of any energy source.
2. Protein corresponds to animal power—horses, oxen, and so forth. Until recent times, humanity used this source very much.
3. Carbohydrates correspond to trees, wood, grass, and paper. Out of these we can create energy.
4. Water is equivalent to waterpower, including mills and hydroelectric dams.
5. Air corresponds to wind power, heat and cold, or in other words,

climate. This includes solar energy, although solar energy is also vibration; it is in between levels five and six.

6. Vibration corresponds to waves: short waves, long waves, and radioactive waves.

7. Electromagnetic forces, or infinite force. This is the force between poles, or yin and yang.

In civilization, if the sources of energy are maintained in the one to seven ratio as described above, then that civilization is technologically accomplished. In this way, the order that governs our bodies, our mind, our entire human nature, can become manifest in civilization as a whole. Modem civilization is using mostly the source that should be utilized the least: deposited resources, such as coal and oil. In ancient times, people were using more of the third, fourth, and fifth levels—wood, water, and wind. But now a great imbalance has arisen, so self-reflection is occurring; people are studying the use of solar power and the possibilities of reintroducing wind and hydropower. But that is not enough; we must restore all of these sources of energy in a balanced ratio of one to seven.

We are using oil and coal for energy, but by themselves, these things do not produce energy. We must burn them. In other words, to obtain energy, we must make these substances expand and decompose. During the process of decomposition, force comes out, which we then use. Coal and oil are condensed: when they decompose, the energy they contain is returning to its original diffuse state. When these substances are burned, they are returned toward heaven, toward the universe.

Our thinking, in the present day, is to obtain energy by using outward movement through the burning or decomposition of deposited resources. But the creation of the universe occurs in exactly the opposite way. Force from the universe is coming in, not going out. Vibration from the infinite universe is constantly coming in to the earth and creating air, water, everything. Man-made, artificially produced energy and universal, natural energy are going in totally opposite directions.

New technologies should depend upon the infinite force coming in to us from near or far: from faraway galaxies, from the sun, and from the depths of the universe. We should be using the energy coming in toward the earth, instead of relying on materials being dug up from the earth. Force going outward is yin; the force coming in is yang—gathering, condensing, and materializing. The earth was created by this more yang force. This is also creating human life. The energy coming in to the earth from heaven is enormously powerful. We must learn how to use whatever is coming in to the earth, including solar wind, short and long waves, and cosmic energy. We must learn to use the force of the universe itself.

When two points—atomic transmutation and the method for securing unlimited energy—are accomplished, the development of technology is complete. This will take many years, but the time will eventually come—especially if people who know yin and yang devote themselves to solving these technological problems—when a new civilization, a new technology, a new orientation will begin on this earth.

The modern way of using energy goes against the creation of the universe, and this has produced many difficulties. Humanity has had to struggle constantly. Money, people, and resources have been required to make these technologies work, and as a result of all this effort, materials are now becoming scarce and the entire planet has been polluted. These unfortunate results have arisen because we have been trying to go against the current of the universe.

If we turn 180 degrees; our progress will become much easier. We have already seen many people with illnesses that modem medicine could not help recover easily and simply through macrobiotics, at very low cost. In the same way, once we change our technological orientation 180 degrees, our path will become much easier.

Technology should be universal and practical, so that everyone can use it at a low price. Anyone should be able to enjoy it. In the same way, everyone should be able, to maintain his or her health in a simple and inexpensive way. Everyone should be able to have heating, lighting, and so forth, easily and cheaply. In the future hopefully we can develop an educational center to display new applications of technology: agricultural technology, food processing technology, housing technology, energy technology, and transmutation. These new technologies should be simple, practical, and economical, yet very effective.

18

TRANSMUTATION OF SOCIETY

The school of Lao Tsu had four stages. First, the mental and psychological aspects of a person's nature had to be mastered. After that was accomplished, the law of the order of the universe was applied. Longevity was mastered. Then these principles were taught and spread throughout the world. The last stage was the understanding of alchemy. After finishing the fourth stage, no matter how many years it took, a student graduated from the school of Lao Tsu.

Today we have a similar school, the School of Ignorance. The first stage is the development of our judgment, our thinking, our mental, psychological, and spiritual aspects. After that we must apply this thinking to our health. We must heal ourselves and help many other people to do the same. In order to graduate from the second stage, you must help at least 300 people. The third stage involves teaching other people—adults, children, anyone—through speaking and writing. You must spread your understanding of happiness, judgment, health, and the order of the universe throughout society.

The fourth stage is the application of the order of the universe to the technological problems of today in the world of matter. You should be able to make matter and utilize it freely. You should be able to produce iron or gold whenever you want them. There our learning is transmutation. You should be able to freely use this matter once it has been produced. In this field the major problem today is energy. The time has come for a student of the order of the universe who has reached the fourth level to answer the social demands for solving the energy crisis, the economic crisis, the industrial crisis, and the environmental crisis. After you do this you will have graduated. You will be a free person.

Those of you who were born in a rich society have a tendency to ignore the economic, industrial, and material problems of the modem world. However, had you been born in a poor society, you would view material civilization through different eyes. But since you were born into it, you would rather decline it, escape from it, and try to go in the opposite direction, for example toward spiritual or mental development. That is the first level of the School of Ignorance: even before we begin to establish our health, even before we begin to chew brown rice, we must master this first stage. It takes a long time. And for many the task is hard. We have to struggle against the modem education that has given us many unnecessary concepts and an upside-down view.

So we have to go back to the beginning, get rid of unnecessary illusions, and become like children, with open and unspoiled minds. We must start to see again nature, the universe, people, society, and find out why, how, what. Then we will begin to understand the order of the universe, and that it is working constantly. We begin to know that whatever has a beginning has an end, that whatever has a front has a back. Once our eyes begin to open to the development of judgment, we can start to apply this to health.

At puberty, we start our own journey of biological history. Before that we depend on our mother, father, and family. Whatever they were giving, we were taking, even though we selected our parents. We do not take the initiative until the time of puberty. By that time our judgment should be developed to the point that we are ready to begin advancing toward the second stage.

At that time you should begin to apply your understanding to acquiring and maintaining health. That is what ancient people referred to as "longevity," or the ability to achieve physical, mental, and spiritual health and wellbeing. The next stage follows from this one. The third stage involves helping society reach a state of good health. At this stage you are able to help people with physical, mental, and spiritual sickness recover their health.

Many of our friends already know how to prevent cancer naturally. If we help many people prevent cancer through our methods, society can save the billions of dollars it spends each year on cancer research and treatment. It costs practically nothing to prevent cancer through a balanced, natural diet. In fact, even the cost of daily food becomes much cheaper. Asthma, mental illness, various kinds of sicknesses can be prevented at very low cost and with very simple measures. Such wonderful things we have seen already. This is the treasure of treasures, just many of our friends have not used this knowledge, even though they can do it. Jesus, at the age of thirty-two, started to use it. Many of us are already older than he was when he began. Why not start to use this knowledge?

As you start to use more of what you know, you start to understand more. Then you go from there to spread, to encourage, helping other people. What is life? What is happiness? What is freedom? What is love? What is justice? You spread and you inspire. That is the third level. You can work by speaking, by writing, or through your example, by practicing and realizing health and peace for yourself. After many years of doing this, you will have many friends. For many years, George Ohsawa worked on this third level. Buddha spent his whole life on the second and third levels.

The fourth level deals with the change of matter, the change of material civilization. Material civilization is the product of human thought, therefore, we should be able to use human thought to change it, especially since the transmutation of the elements has added an entirely new dimension to our understanding. Although there are many varied aspects to industry, economy, and society, there are basically only two problems. The first is how to obtain materials, or matter. At the present time we are taking raw materials from the earth, by mining it, or from the ocean. But if instead of mining these raw materials, if we could make them simply and freely from the air, from the water, or from simple beach sand, then we would not need to pollute the environment or deplete the earth's natural resources.

That brings us to the second problem. After we have made raw materials, then we must be able to use them. For that we need energy. Today energy is obtained from deposited sources such as oil, coal, and natural gas. Even nuclear power comes from mined natural resources. But suppose we were able to produce energy from the air, from the wind, and from the electromagnetic force that is constantly coming from the universe, then we would not have the problem of pollution, the problem of huge power plants, or the problem of diminishing natural resources. These two points—how to make matter, and how to invent energy—are the key not only to solving the environmental crisis, but also to transforming modem industrial society.

For students in the School of Ignorance, the first stage is the most difficult. The most difficult step to undertake is the development of the understanding of yin and yang, of the training of your own will and judgment. The second most difficult step is the healing of sickness. Much easier is the dissemination of the principles you have acquired for the benefit of society. The easiest step is the transmutation of matter. And that is the stage we are now reaching at this time.

Spiritual development is on the first level. It is at the level in which we work on developing our judgment. The conclusion of that process is the awareness of yin and yang, or the unchanging law of the infinite universe. However, please develop further than this. Challenge and apply your understanding to newer and larger dimensions, progressing from self-development to the change of society and the whole of material civilization. Simply trying to escape from civilization is a negative response. It is far more positive and challenging to change the orientation, values, and techniques of civilization as a whole. The transmutation of the atom is a pivotal step in the transmutation of society toward peace, harmony with the environment, and the ability to play freely with matter and energy.

Afterword

The takeoff from Hartford was smooth and trouble-free. The USAir 737 turned out over the Atlantic and headed south. The weather was rainy and the temperature mild for early December. After a brief stop in Charlotte, I boarded the forty-minute connecting flight to Atlanta. This was my third visit to Atlanta for lectures since 1984. Following Atlanta, I was scheduled to return to Charlotte for several days of teaching.

During the flight, I reflected on the book I just finished editing. Titled *The Philosopher's Stone,* it is based on Michio Kushi's lectures on alchemy and transmutation. According to legend, the philosopher's stone was the mysterious element used by medieval alchemists to transmute base metals into gold. To me, the philosopher's stone symbolizes the invisible law that produces all of the changes in the universe, including the transmutation of one element into another.

Transmutation has intrigued me since the beginning of my macrobiotic practice over twenty years ago. Much of the early macrobiotic literature contained references to the work of Louis Kervran, George Ohsawa, and others in the field of transmutation. Kervran, a French biochemist, discovered the transmutation of sodium into potassium in French workers in the Sahara. His findings were summarized in the book, *Biological Transmutations* (Happiness Press, 1987). George Ohsawa worked with Kervran and devoted the later years of his life to proving transmutation in the laboratory. In my thinking, transmutation offers proof of the mutability of the material world, and is at the core of the macrobiotic philosophy of change. Writing in *Biological Transmutations*, Kervran describes the relationship between transmutation and modern chemistry and physics:

The serious error of scientists consists in their saying that reactions occurring in living matter are solely chemical reactions, that chemistry can and must explain life. That is why in science we find such terms as "biochemistry." It is certain that a great number of manifestations of life are produced by chemical reactions. But the belief that there is only chemical reaction and that every observation must be explained in terms of a chemical reaction, is false. One of the purposes of this book is to show that matter has a property heretofore unseen, a property which is neither chemistry nor nuclear physics in its present state. In other words, the laws of chemistry are not on trial here. The error of numerous chemists and biochemists lies in their desire to apply the laws of chemistry at any cost, with unverified assertions, in a field where chemistry is not always applicable. In the final phase the result might be "chemistry," but only as a consequence of the unperceived phenomenon of transmutation.

When I visited Prague in 1990, friends took me to a section of the city called "Alchemists' Row." It is a narrow street on either side of which are curious tiny houses. Our guide explained that these houses were where medieval alchemists had lived and conducted their experiments. Prague was one of the centers of medieval European alchemy. The tiny houses on Alchemists' Row now serve as boutiques and gift shops for tourists.

Over the years, Michio Kushi has lectured on transmutation, and transcripts of these lectures were published in his seminar reports. Michio worked with Ohsawa and Kervran on transmutation experiments in New York and Cambridge. When I edited Michio's book *Other Dimensions: Exploring the Unexplained* (Avery, 1992), I included a chapter on transmutation. I also lectured on the macrobiotic view of transmutation at the East West Foundation in Boston, and later as a part of the Kushi Institute's Level III program in Becket.

More recently, Michio announced that a group of scientists at a university in Texas had achieved the transmutation of carbon into iron, using Ohsawa's pioneering experiments as a guide. Michio announced that he was planning to introduce the transmutation of carbon into iron to industry in the hope of perfecting a method for the mass production of steel. As he states in *The Philosopher's Stone*, transmutation could be the key to changing industrial civilization and solving the global environmental crisis.

As the plane landed in Atlanta, my thoughts came back to the task at hand. My hosts had arranged for the seminars to take place at the Doubletree Hotel outside Atlanta. Fifty people came to the Friday and Saturday night lectures. Following the weekend in Atlanta, I returned to Charlotte. Michel Matsuda met me at the airport. I first met Michel twenty years before in Boston, when both of us were studying with Michio and living in a macrobiotic student house. When Wendy and I visited Japan in 1978, Michel helped us get settled in Kyoto, his home city. Michel and his wife Libby, who is originally from Ireland, now run a macrobiotic study center in Charlotte. Aside from being a student of macrobiotics for the past thirty-five years, Michel is a skilled acupuncturist. His practice in Charlotte is now quite active.

After leaving the airport, Michel and I went to a Japanese restaurant not far from his home. I mentioned the recent developments with atomic transmutation. Michel told me that he had been involved in transmutation research in Kyoto in the mid-Sixties. He led a group of young students, known as the "Circle of Seven" (named after the Kurosawa film *Seven Samurai*) in the study and practice of transmutation. The group, inspired by the work of George Ohsawa and his associates, met weekly in an abandoned textile warehouse in Kyoto. The group started out with seven members, and eventually grew to include several dozen.

In June 1964, Ohsawa achieved the transmutation of sodium into potassium (with the addition of oxygen) in the laboratory, under low temperature, pressure, and energy. He later achieved the transmutation of carbon into iron (with the addition of oxygen) under similar conditions. These discoveries challenge the prevailing notion that elements are fixed and separate, and change into each other only under very extreme circumstances, such as in a particle accelerator, during a thermonuclear reaction, or in the sun.

Inspired by Mr. Ohsawa's results, the Circle of Seven met weekly from 1965 to 1967. According to Michel, after much trial and error, they achieved the low-energy transmutation of carbon into iron. For Michel, these times were the most exciting in his thirty-five years of macrobiotic practice. He gave me the addresses of several member of the Circle of

Seven after I expressed interest in writing to them for more information about their experiences. In 1978, the U.S. Army commissioned a report on biological transmutations. The report concluded:

> Two investigators, Kervran and Komaki [an associate of George Ohsawa's], have been recently nominated for a joint Nobel prize for their work involving experimental proof that elemental transmutations were occurring in life organisms. Elements which were definitely proven to have been transformed were sodium (to magnesium), potassium (to calcium), and manganese (to iron). Actually, observations have been made for almost 200 years that elemental transmutations were occurring, but little credence was given to them because they resembled alchemy—a relic of the Middle Ages.

Modern physics and chemistry were born in the laboratories of the medieval alchemists. However, the quest for the philosopher's stone took a destructive turn in the 20th century. In place of the peaceful, natural methods employed by ancient alchemists, modern researchers began to utilize violent and destructive methods to achieve transmutation. In 1920, Rutherford changed nitrogen into hydrogen and oxygen by bombarding nitrogen atoms with subatomic alpha particles. Ten years later, Ernest Lawrence invented a device called the cyclotron in which atomic particles were accelerated with high energy and used to "smash" target atoms. In 1932, scientists discovered that neutrons could be used as "bullets" to smash atoms, and in 1939, the nucleus of uranium was "bombarded" with free neutrons, causing it to "split" and release energy. In 1942, Enrico Fermi at the University of Chicago used these discoveries to achieve a chain reaction, and soon afterward, Oppenheimer and other researchers in the Manhattan Project used these discoveries to build the first atomic bomb.

After World War II, scientists used these discoveries to pursue nuclear fusion. Using an atomic bomb, they forced two atoms of hydrogen to fuse and form an atom of helium, releasing tremendous energy in the process. That led to the development of the thermonuclear, or hydrogen bomb. In 1953, the United States and the Soviet Union began to actively manufacture these weapons of mass destruction. Since then, nuclear weapons technology has spread around the globe. According to *Newsweek*, twenty-five nations have, or may soon have, nuclear weapons. The disposal of nuclear waste is also a gigantic problem. The U.S. Government recently revealed that it stores 33.5 metric tons of deadly radioactive plutonium in six states. Many of the storage facilities for nuclear waste are old and deteriorating rapidly. As we can see, the modern scientific pursuit of the ancient dream of the alchemists has led to a situation that threatens both humanity and the environment.

The discovery of peaceful, natural transmutation offers an alternative to these destructive methods. Atomic transmutation can be achieved under natural conditions without having to attack and destroy atoms. The work of George Ohsawa, Louis Kervran, Michio Kushi, the members of the Circle of Seven, and other pioneers in peaceful, natural transmutation have shown that the world of matter is not fixed and static, but dynamic and changing.

These discoveries could revolutionize science and open the door to a new era for humanity. If the knowledge of transmutation is properly understood and applied, the age-old quest for the philosopher's stone will contribute to an age of peace and prosperity. The transmutation of the atom is thus a metaphor for the transmutation of society itself.

Source: This essay is from an article entitled, "Reflections on the Philosopher's Stone," published in *MacroNews*, Philadelphia, PA, Spring 1994 and reprinted in *Contemporary Macrobiotics*, Amberwaves Press, 2012.

PART II

Cool Fusion

The experiments described here are done in simple laboratories
with a low budget, yet are significant because they show with
little doubt that elements can be transmuted at temperatures
far below what science would consider possible.
BILL ZEBUHR, INFINITE ENERGY MAGAZINE

FOREWORD

Edward Esko and Alex Jack have continued the legacy of the pioneers George Ohsawa, Louis Kervran and Michio Kushi of the field now widely known as "Low Energy Nuclear Reactions" (LENR). I first heard about Kervran's book *Biological Transmutations* through Eugene Mallove in the early '90s and was fascinated by the concept of occurrence of nuclear reactions in living organisms. We even carried out some experiments on the possible occurrence of elemental transmutations during germination of seeds at BARC in the mid-'90s. Although we did get what appeared to be preliminary positive results, the effort did not lead to a publication as I retired from BARC shortly thereafter. However Vladimir Vysotskii of Ukraine has during the last decade carried forward the field further, building upon the foundation laid by Kervran.

During my tenure at BARC I did have occasion to encourage some of my colleagues in the spectroscopy division to verify Ohsawa's carbon arc experiments, which were simultaneously, also replicated at Texas A & M University, since one of our colleagues, Dr.Sundaresan was doing a post doc with Prof. Bockris at that time. Both these carbon arc papers were jointly peer reviewed and published in the same issue (November 1994) of *Fusion Technology*, courtesy of George Miley, its then editor, who even wrote a special editorial covering these two papers!

Much progress has taken place since those days in establishing firmly the occurrence of different types of transmutation reactions in a wide variety of LENR configurations. A review article on the experimental findings, written jointly with Edmund Storms and George Miley has appeared in the 2011 edition of the *Wiley Encyclopedia on Nuclear Energy*.

The year 2011 also witnessed great excitement in what appears to be industrial level heat production in Ni-H gas loaded LENR "reactors" wherein the inventors claim that the excess heat is caused by a nuclear reaction involving the transmutation of nickel isotopes to those of copper. However this claim has yet to be quantitatively established through reliable analysis of post run spent-Ni fuel powder samples.

During the last six years (since 2005) Edward Esko and Alex Jack of the Quantum Rabbit company have been quietly carrying out a series of systematic experiments with evacuated electrical discharge tubes designed to verify preselected transmutation reactions, much as what George Ohsawa did almost half a century ago (1964 to be precise). The Quantum Rabbit results were initially summarized in several articles in *Infinite Energy* magazine during the last few years and last year compiled in the form of the first edition of their monograph titled *Cool Fusion*. Their results certainly appear very promising and deserve to be independently confirmed, preferably in a proper academic setting. As conceded by the experimenters themselves, all suspicion of contamination being the cause of the apparent transmutation observations needs to be ruled out.

In this context the very bold initiative launched by Lewis Larsen of Lattice Energy LLC very recently is worth taking note of. At a time when the global price of gold is hitting the roof, Larsen has announced his interest in embarking on a semi commercial venture to produce gold using LENR devices starting from inexpensive tantalum, even inviting collaborating partners!

Lattice Energy's confidence in such an approach is based on their weak interaction theoretical model superposed on all the experimental results cited above (although other theoreticians have expressed some reservations regarding their model). I draw attention to the Larsen initiative only to highlight the farsighted vision displayed by the authors of *Cool Fusion* who had undertaken their experimental studies keeping in mind a time not too distant into the future when mankind may run out of precious heavy element metals. Their motivation in undertaking such experimental investigations was to pave the way in establishing a technology, which can be relied upon to hopefully come to the rescue of mankind when such an eventuality arises!

An added incentive is the potential application of LENR transmutation technology to convert radioactive waste left behind by the nuclear fission power industry into more benign stable elements!

Those of us who have been watching the cold fusion-LENR drama unfold from a "ring side seat" over the last couple of decades have been witness to the many twists and turns it has taken. We however feel fortunate to be part of scientific history being made. It is hoped that this second edition of *Cool Fusion* will serve as an inspiration to newcomers to the field and embolden them to embark on an exciting scientific journey, with a view to placing Low Energy Nuclear Transmutations on an even more strong scientific footing! Good Luck Edward and Alex!

Mahadeva Srinivasan
Bhabha Atomic Research Centre (BARC)
Mumbai (Retired)
27th May 2012

Introduction

Edward Esko addresses the International Conference on Future Energy

On the drive back from the Third Conference on Future Energy in Washington, DC, Florence Johnson hands me the latest issue of *Macrobiotics Today*. She points to the article, "George Ohsawa's Last Letter—To his students in the United States." It is October 2009 and, with Woody Johnson at the wheel of his hybrid SUV, we are heading back to Massachusetts after making a presentation at the Washington DC Conference. It is a fine sunny day along the Eastern Seaboard.

In his "Last Letter," Ohsawa states, "On New Years Day, 1964, I began the final course of Lao Tsu: alchemy. Surprisingly, I finished it very rapidly, for as soon as I had gathered the equipment necessary for experimentation, I achieved the transmutation of sodium to potassium (Na to K). The date was June 21, 1964—only five months later. "Upon my return from a world trip, I began experimenting with the transmutation of carbon into iron (C into Fe). This was also successful. Is it not wonderful that when I asked that sodium change to potassium, it happened? And when I asked that carbon change to iron, it was no sooner said than done? By November I was able to conclude that all elements up to atomic number 82 (lead) could be transmuted from lighter elements like carbon, oxygen, or lithium. This achievement is proof of what can be realized through a deep understanding of one of the fundamentals of the macrobiotic principle: Nothing is eternal—everything changes in this relative world."

Ohsawa died a year after writing this letter. He didn't have the chance to develop these discoveries beyond the initial stage. Had he had another ten years to develop his transmutation work, the world we inhabit today would be vastly different. We would be much nearer to a sustainable future based on universal health and prosperity. For me, reading Ohsawa's letter in 2009 could not have been timelier. I had just given a PowerPoint presentation in which I described five low energy nuclear reactions (transmutations) achieved at our Quantum Rabbit lab. Quantum Rabbit is in essence an advanced study group that I started together with Alex Jack and Woody Johnson. Its purpose is to explore and develop new applications of yin and yang (yin/yang apps) in a variety of domains, beginning with the world of elements and energy.

After more than thirty years studying how to use the vegetable world in diet, health, and healing, including the health of society and the planet as a whole, I decided it was time to move on to the next level of study: the world of elements. Studies began in earnest early in the year 2000, even though I had lectured on Ohsawa's research beginning in 1975 at the East West Foundation and later at the Kushi Institute. In 1994 I edited Michio Kushi's lectures on this topic and added original insights. One Peaceful World Press published this as a small book, *The Philosopher's Stone*.

As Michio Kushi once said, after you reach the top of the mountain, you look out and see an even higher mountain in the distance. That higher mountain is the seemingly fixed and impenetrable barrier known as the elemental world. Unlike the soft, feminine world of plants, which readily yield to the cut of the knife and fire of the stove, the solidly masculine world of elements appears as hard and unchanging as a stone on the beach, the imposing rock faces on Mt. Rushmore, and the brilliant 24-carat diamond in an engagement ring. As the saying goes, "diamonds are forever."

Although it is acknowledged that atoms are composed primarily of empty space, with no solid substance, the periodic table gives no indication that elements can in any way change into one another. On the surface at least, the elements seem to occupy a fixed position in a universe that is static and unchanging. It was this world that we decided to challenge. It was this barrier that we sought to breach so as to enter a new, quantum reality unbound by artificial constructs and governed only by the endless law of change. We would work and become familiar with the chemical elements just as we had previously worked and became familiar with brown rice, carrots, barley, sea salt, and cabbage in our kitchen workshops. This time, our workshop would take the form of a small laboratory with vacuum equipment, an electric power supply, and pure element samples.

The origins of alchemy are shrouded in myth and legend. Ancient Egyptian writings describe "visitors from the firmament" who shared their knowledge of the universe with humanity, including the practice of alchemy. Supposedly, godlike beings arrived in Egypt who possessed an advanced spiritual technology through which they were able to transform matter. Medieval alchemists are the forefathers of modern chemists. Alchemy was practiced in ancient Greece, Arabia, Europe, India, and China, in addition to its legendary origins in Egypt. More recently, we trace our lineage to Sir Norman Lockyer, founder of the British publication *Nature* and discoverer of the element helium. An article published in the *New York Times* on December 13, 1878 describes Sir Norman's exploits. The article, titled "Transmutation of Metals—Is the Old Dream of the Alchemists to Be Realized?" talks about his curious results:

A correspondent from the *London Daily News* writes: "Today, in the presence of a small party of scientific men, Mr. Lockyer, by the aid of a powerful voltaic current, volatized copper within a glass tube, dissolved the deposit formed within the tube in hydrochloric acid, and then showed by means of the spectroscope, that the solution contained no longer copper, but another metal, calcium, the base of ordinary lime. The experiment was repeated with other metals and with corresponding results. Nickel was thus changed into cobalt, and calcium into strontium. All these bodies, as is well known, have ever been regarded as elementary, that is, incapable of being resolved into any components, or being changed into one another. It is on this basis that all modern chemistry is founded, and should Mr. Lockyer's discovery bear the test of further trial, our entire system of chemistry will require revision. The future possibilities of the discoveries it is difficult to limit...."

"Mr. Lockyer is one of our best living spectroscopists, and no man with a reputation such as his would risk the publication of so startling a fact as he has just announced to the scientific world without the very surest grounds. He was supported yesterday by some of our leading chemists, all of whom admitted that the results of his experiments were inexplicable on any other grounds but those admitting of the change of one element into another..."

Apparently, Sir Norman achieved his transmutations using a very basic method, without super-high temperatures, pressures, or energy. Unfortunately, nothing seems to have come from these experiments. Meanwhile science was busy pursuing another approach to transmutation, an approach with potentially sinister and destructive implications: the bombardment of atoms with radioactive particles. In 1907, Sir William Ramsay, one of England's top chemists, announced that he had achieved the transmutation of elements using the radioactive "emanations" of radium. His announcement was published in the *New York Times*, July 27, 1907, in an article titled, "Turns Copper into Lithium," with the subtitles: "Sir William Ramsay Effects the Transmutation of the Elements Sought by the Alchemists—Change Caused by Radium, Theory that the Elements of High Atomic Weight will Disappear—Statement by Ramsay."

LONDON, July 26—Sir William Ramsay has promised to communicate shortly to the Chemical Society an account of a discovery, which, in the words of so conservative a scientific publication as *The Lancet* in its number issued today "marks an epoch in the history of chemical science, since his investigations have shown that a given element under the powerful action of radium emanations undergoes degradation into another. In short," adds *The Lancet*, "the transmutation of the elements is actually un, fait accompli." Reversing the process sought by ancient alchemists, who believed that there was a substance by means of which the baser metals could be transmuted into the higher, Sir William has effected the degradation of metals by means of gas evolved from radium. The paper will prove that Sir William has degraded copper to the first member of its family, namely, lithium. In other words, he has effected the transmutation of copper. Commenting on Sir William's experiments, *The Lancet* continues: "These remarkable discoveries remind us again of the extraordinary prescience of the ancients and of the presentiments of the alchemists, who evidently had some sort of conviction that after all there is a primary matter from which all other elements are formed by various condensations. He is a bold man who nowadays confesses skepticism about anything. The world has seen men who have said 'it is impossible,' and generations who succeeded them who have seen the impossible come true."

In the thirty years between Sir Norman's remarkable experiments and Sir William's demonstrations, the discovery of radioactivity by Marie Curie in 1897 would prove decisive, especially the radioactive properties of uranium. It was these radioactive "emanations," specifically those of radium that enabled Sir William to achieve his novel results. From then on, alchemy took what I consider to be a decidedly wrong turn, transforming itself from what I call "golden" alchemy, or the pursuit of transmutation using relatively natural (low energy, temperature, and pressure) conditions for the benefit of humanity, to what I term "dark" alchemy.

Dark alchemy utilizes dangerous radioactive elements to produce nuclear reactions (fission and fusion) through the deployment of highly destructive, expensive, dangerous, and wasteful technologies with potentially catastrophic consequences. The discovery of the radioactive properties of uranium, one of the extremely heavy elements at the far end of the periodic spectrum, was the driving force behind these developments. Today's arsenals of atomic and hydrogen bombs are the legacy of this latter approach, as is the proliferation of atomic weapons and ever-increasing stockpiles of plutonium and other highly toxic materials.

It was in the midst of the nuclear nightmare spawned by 20th century physics that the first rays of purifying sunlight began to appear in the person of Louis Kervran. Kervran appeared out of nowhere like the proverbial lotus flowers emerging from the proverbial mud. Kervran was the tiny dot of Yin within the larger Yang; the tiny Yang within the all-encompassing Yin. Although small and seemingly insignificant, the contrary dots contain the seeds of great change. They are harbingers of things to come, like the tiny mammals that scurried to avoid being trampled by the giant dinosaurs. Through his discovery of biological transmutations, Kervran predicted 21st century physics and technology, in which science would accept and develop golden alchemy, triggering a paradigm shift capable of averting further human and ecological catastrophe. Through years of patient observation, Kervran, a biochemist from Brittany in France, discovered that transmutation occurs naturally, at no cost, and quite peacefully in the life cycle of plants and animals. Kervran's formulas were an organic, refreshing, and much needed antidote to cyclotrons, aboveground hydrogen bomb tests, and nuclear reactors. It had finally come down to a choice between organic alfalfa sprouts or reactor waste.

Kervran's well-known formula, $^{23}_{11}Na + ^{16}_{8}O \rightarrow ^{39}_{19}K$ (sodium-23 + oxygen-16 into potassium-39), a process he observed in the human body, would become the cornerstone of Ohsawa's experiments. It would also serve as the basis for Quantum Rabbit low energy formulas, including several in which a highly toxic radioactive element such as plutonium can be fractioned into a benign non-radioactive element such as bismuth or lead. When I spoke in August 2008 at the Transmutation Workshop organized by Prof. George Miley (a nuclear fusion expert from the University of Illinois) at the International Conference on Cold Fusion (ICCF-14) held in Washington, DC, I asked the group of about fifty international scientists if they knew of Louis Kervran. Practically everyone raised his or her hand. When asked, a surprising number also knew about Ohsawa's transmutation work.

Aside from the mini-revival of carbon-arc studies in the 1990s at Texas A & M University, in India, Japan, and by Michio Kushi and Chris Akbar in Brookline, inspired by Ohsawa (and by publication of *The Philosopher's Stone*); and *all* of which confirmed Ohsawa's earlier findings, I had assumed that research on low energy transmutation had come to a standstill due to inertia and indifference on the part of the scientific community. As it turned out, I was wrong. The first indication of this came when Alex, Woody, and I went to the Massachusetts Institute of Technology (MIT) in September 2007 to present the results of our carbon-arc experiments.

Armed with PowerPoint slides, video clips, lab reports, and formulas we entered the citadel of modern chemistry and physics to meet with MIT Professor Dr. Peter Hagelstein, one of a handful of mainstream scientists who have continued with research on cold fusion following the highly controversial 1989 disclosure of this potential source of free energy. Although the focus of cold fusion is on the creation of energy, a number of investigators have noted the creation of new elements during the process, adding possible confirmation to the theory of low energy transmutation. Dr. Hagelstein listened to our presentation and offered valuable suggestions, including adding high-energy particle detection slides to our studies, to see if the transmutation process produces energy.

Interestingly, in our carbon-arc studies, based on those of Ohsawa, we discovered not only the possibility of transmuting carbon into iron (with oxygen from the atmosphere), but also a host of other reactions involving *both* carbon (yang) and oxygen (yin) and carbon (yang) and nitrogen (yin) from the surrounding air. Our pure graphite (carbon) test samples showed, in addition to iron, the presence of magnesium, aluminum, silicon, scandium, titanium, cobalt, and nickel—in what appeared to be a veritable cascade of transmutations. We reported these to Dr. Hagelstein. We also showed clips in which we tested graphite powder for magnetic properties. Using powerful magnets composed of neodymium, one of the rare earth metals, all of the treated samples showed magnetic activity, while untreated samples did not.

Dr. Hagelstein told us of other work being conducted on cold fusion and transmutation around the world. He mentioned the work done at Mitsubishi

Heavy Industries in Japan in which researchers reported the ability to transmute one heavy element into another, notably cesium (Cs) into praseodymium (Pr) and strontium (Sr) into molybdenum (Mo). Fifteen laboratories in six countries have reported evidence for transmutation. Isotope ratios that deviate from natural abundances have also been reported. Such variation in the isotope distribution of an element is considered evidence it may have formed through transmutation rather than being present beforehand as a contaminant.

Our vacuum studies have been conducted at small labs in Nashua, New Hampshire and Owls Head, Maine at Rockland Harbor. We had to more or less start from scratch, reinventing the wheel so to speak, as details of the Ohsawa experiments were sketchy at best. We were entering uncharted territory, out on the open sea with only the compass of the unifying principle to guide our direction. Fortunately, we teamed up with one of the top vacuum consultants on the East Coast, along with a master scientific glassblower who fabricated beautiful and functional vacuum tubes according to my design specifications.

We started initially with experiments on the noble gases, helium (He), neon (Ne), argon (Ar), and krypton (Kr). After much trial and error, we achieved what appeared to be the transmutation of helium (He) into argon (Ar), with the addition of oxygen (O_2), with the applied formula: $^4_2He + _2(^{16}_8O) \rightarrow {}^{36}_{18}Ar$. This formula seemed to confirm Ohsawa's hypothesis that a yang element (helium) would readily fuse with a yin element (oxygen) to form a new, hybrid element, argon.

From the noble gases we moved on to research with metals, including several attempts to duplicate Ohsawa's sodium into potassium experiment. We failed repeatedly. Undaunted, we persevered in our studies. It was not until May of 2008 that we stumbled upon an unexpected roadmap to guide us in our research. The key was again the relationship between sodium and potassium. If you look at the periodic table, you see that the element sodium (atomic number 11) and potassium (atomic number 19) are separated by an element with the atomic number 8. That element turns out to be oxygen $(^{16}_8O)$. In other words, it is an atom of oxygen that separates sodium from potassium, as Kervran discovered and Ohsawa demonstrated.

Alex Jack (left) and Woody Johnson at the Nashua lab

Looking further at the periodic table, we see that sodium and lithium have a similar relationship. In other words, lithium, with the atomic number 3 and sodium, with atomic number 11, are also separated by atomic number 8; once again, an atom of oxygen. These three metals, lithium, sodium, and potassium are related; they share many characteristics as members of the Group IA elements known as the Alkali Metals, for example low melting temperatures, light weight, softness (you can cut them with a knife), and extreme volatility. Sodium, for example, will explode if placed in water.

Using Ohsawa's principle of change, we see that lithium is actually the precursor to sodium, and sodium the precursor to potassium, each separated by oxygen. With this background I designed an experiment in which we would try to *quantum bounce* lithium further along the periodic table, so that it changed into sodium and then to potassium, according to the applied formula: $^{7}_{3}Li + {}^{16}_{8}O \rightarrow {}^{23}_{11}Na$; $^{23}_{11}Na + {}^{16}_{8}O \rightarrow {}^{39}_{19}K$ (lithium-7 + oxygen-16 $\rightarrow$ sodium-23; newly formed sodium-23 + oxygen-16 $\rightarrow$ potassium-39.)

Another variation of the reaction is: $^{7}_{3}Li + {}_{2}({}^{16}_{8}O) \rightarrow {}^{39}_{19}K$, or lithium-7 fuses with two atoms of oxygen-16 (a molecule of oxygen, or O_2), skipping sodium-23 and *quantum bouncing* to potassium-39. I devised two experiments to test this hypothesis. In the first two, conducted on February 29 and May 2, 2008, we used pure lithium metal, stainless electrodes, and pure oxygen as fill gas in the vacuum tube.

All the test samples showed the presence of sodium (at up to 0.94%) and potassium (at up to 0.14%.) We were excited, thinking we had proven the formula. I wrote a letter to *Macrobiotics Today* announcing this result. However, upon closer examination, we discovered that the borosilicate (Pyrex) glass used in the vacuum tube contained a significant trace of sodium and minor trace of potassium, enough to influence the outcome of the experiment. I scheduled another series of tests for May 30 with the goal being to control for sodium. A new tube was designed so that the reaction zone at the center would be made of quartz. The quartz contained less than one part per-million sodium, so we were confident it wouldn't influence the outcome.

After conducting the test, we noticed that the tips of the stainless steel electrodes had undergone a color change, with a copper-colored residue on the surface. We sent the materials off to the outside lab for third-party analysis. When the results came back, we were initially disappointed. There were no detectable traces of sodium or potassium, although copper was present, for no apparent reason, on one of the stainless electrodes. I remember Alex commenting somewhat dejectedly, "There's nothing there." It took several days for us to realize that we had achieved an unexpected result: the presence of a significant trace of copper on the surface of the stainless electrode. It suddenly dawned on me what we had done. We had apparently caused lithium (Li-7) to fuse with the iron (Fe-56) in the stainless steel electrode to produce copper (Cu-63), according to the applied formula: $^{7}_{3}\mathrm{Li} + {}^{56}_{26}\mathrm{Fe} \rightarrow {}^{63}_{29}\mathrm{Cu}$.

I realized that we could now use metallic lithium (3), the first solid element, following hydrogen (1) and helium (2) on the periodic table, as the catalyst or trigger for a wide range of low energy transmutations. Interestingly, when lithium metal is vaporized in the vacuum tube, it emits a deep ruby red glow, which is apparently yang. Classical alchemists described the Philosopher's Stone, the mysterious substance used to transmute the elements, as having a ruby red color. Coincidence?

I set out to test the lithium hypothesis in several additional tests. In one, I predicted that lithium-7 would fuse with silver-109 to form tin-116. In another, that copper-63 would fuse with lithium-7 to produce germanium-70. These formulas were confirmed in the autumn of 2008. As predicted, tin (atomic number 50) appeared on the surface of the silver anode (atomic number 47) following a low energy nuclear reaction with lithium (atomic number 3). Germanium (Ge) has the atomic number 32. It has repeatedly appeared in tests in which copper electrodes (atomic number 29) were used with lithium metal (atomic number 3).

Our success with lithium prompted me to search for another solid catalyst. Sulfur (S) turned out to be the perfect candidate. In one test, we were able to fuse both lithium and sodium with sulfur to produce potassium, a variation of Ohsawa's sodium to potassium experiment. We were able to produce potassium-39 from the fusion of lithium-7 and sulfur-32. In addition, sulfur may have fissioned into two atoms of oxygen-16, each of which fused with an atom of sodium -23 to form potassium-39. In our view, sulfur-32 is formed by the fusion of two atoms of oxygen-16. Sulfur is actually crystallized oxygen. In another test, we used a zinc (atomic number 30) electrode in combination with sulfur (atomic number 16) and oxygen test material. As predicted, palladium (atomic number 46), a rare member of the platinum group of metals, was found after the test. I had predicted this with the formula: zinc-68 + sulfur-34 → palladium-102.

Many discoveries lie ahead. Our work so far on transmutation has been unsophisticated and tentative, kindergarten at best. Whether these results lead to a Golden Age of peace and prosperity remains to be seen. However, we are confident that our studies, together with those of our colleagues around the world, shall establish once and for all Ohsawa's most fundamental principle: "Nothing is eternal. Everything [and most certainly the chemical elements] changes in this relative world." That much is certain. Ohsawa concluded his letter by stating, "In closing, let me sincerely urge you to study even more deeply than ever before." I can answer Ohsawa by stating: "George we will try our best. We know that the unifying principle, which you taught for more than fifty years, holds the key to health, peace, and happiness on this earth. In our time we will do our best to develop the unifying principle in the areas we deem most vital to human health and happiness. It is this dream we hope to pass on to future generations. It is my hope that this small book will help move us in that direction.

Edward Esko
Pittsfield and Lenox, Massachusetts
June 2011

1

APPEARANCE OF SILICON AND METALS IN PURE GRAPHITE

ABSTRACT

Researchers at Quantum Rabbit LLC (QR) in the USA have repeatedly seen the appearance of silicon and a variety of metals in a pure graphite matrix. In a series of studies conducted since November 2006 at the QR lab in Bellows Falls, Vermont, treated graphite powder shows permanent magnetic activity plus the presence of metals in the parts per million ranges.

METHOD

Non-metallic graphite powders (scientific grade 99.999% pure) are placed in a pure (99.999%) graphite crucible. The powders are charged with 36 volts of direct current through a pure (99.999%) graphite rod. The crucible is connected to the negative pole, the rod to the positive pole of a power pack consisting of three 12-Volt solar-charged batteries. The powders receive between 100 to 200 strikes from the charged rod. Upon cooling the powders are tested for magnetic properties with a neodymium magnet before packaging and shipping to an outside lab for EDS and ICP analysis.

RESULTS

The powders display apparent magnetic activity following the above treatment. Moreover, magnetic activity remains in the powders six months after treatment, suggesting the effect is permanent. Treated graphite shows the presence of magnetic iron at a level of up to 1.6% by weight. A typical sample (ICP analysis by New Hampshire Materials Laboratory, August 9, 2007) shows the appearance of silicon and metals in treated graphite as follows:

Element	**Composition Sample (ppm*)**
Silicon	10,500
Magnesium	1800
Iron	4700
Copper	4200
Aluminum	7800
Titanium	440
Sulfur	580
Potassium	1000

*Parts per million

Please note the presence of silicon at 1.5% in the treated graphite. No silicon was used in the graphite materials (rod, crucible, powder) employed in the test. In addition to changes in the composition of the graphite powder used in the tests, changes have been noted in the pure graphite rods used in the experiments. In a study conducted in October 2007, a shiny metallic "bubble" appeared on the striking surface of the rod. Upon analysis, the rod was found to contain the following metals:

Element	**Composition Sample (ppm*)**
Scandium	35
Iron	640
Cobalt	160
Nickel	1120

*Parts per million

CONCLUSION

Quantum Rabbit research has repeatedly demonstrated that carbon-based materials (pure graphite powder and rods) can develop magnetic properties when exposed to carbon arcing. Consistent presence of iron and other metals in treated graphite suggest the possibility that the charging process generates low-energy nuclear reactions (transmutations) that result in the appearance of new elements. The sudden charge of electricity may temporarily neutralize the mutual repulsive force existing between two plus charged nuclei. This instantaneous breach may allow the centripetal Casimir force generated by the vacuum/ether to force the nuclei to fuse and form a heavier atom.

The formulas presented below describe possible low-energy fusion reactions that could explain the presence of the new elements in the treated graphite.

Possible Low Energy Nuclear Reactions in Treated Graphite*

Magnesium-24

$$^{12}C + {}^{12}C \rightarrow {}^{24}Mg$$

Calcium-40

$$^{24}Mg + {}^{16}O \rightarrow {}^{40}Ca$$

Aluminum-27

$$^{12}C + {}^{15}N \rightarrow {}^{27}Al$$

Titanium-44

$$^{28}Si + {}^{16}O \rightarrow {}^{44}Ti$$

Silicon-28

$$^{12}C + {}^{16}O \rightarrow {}^{28}Si$$

Iron-56

$$_2(^{12}C + {}^{16}O) \rightarrow {}^{56}Fe \ ^{(+ \, 2 \, protones)}$$

Potassium-39

$$^{26}Mg + {}^{14}N \rightarrow {}^{39}K$$

Scandium-45

$$^{30}Si + {}^{15}N \rightarrow {}^{45}Sc$$

*Please note that the gases involved in these reactions, oxygen (O) and nitrogen (N), are from the atmosphere.

Our research confirms earlier studies conducted by George Ohsawa and associates and reported in *Infinite Energy*. Further research under rigorously controlled conditions, is needed to determine whether or not low-energy nuclear reactions are taking place in treated graphite materials.

Charging graphite powder with graphite rod and crucible

Source: Edward Esko, "Production of Metals in Non-Metallic Graphite," *Infinite Energy* Issue 78, 2008.

2

APPEARANCE OF ARGON IN OXYGEN-HELIUM PLASMA

ABSTRACT

In gas studies conducted at Quantum Rabbit (QR) lab in Nashua, New Hampshire, USA, in November and December 2005, QR researchers recorded the anomalous appearance of argon in vacuum tubes containing oxygen and helium plasma. Results were monitored and recorded by residual gas analyzer (RGA).

METHOD

Experiment 1: Oxygen plus Helium

4 parts O + 1 part He
Pressure O_2 = 2x He

In the first experiment, the QR vacuum tube was filled with pure oxygen, isolated, and pumped down to approximately 4 torr. Plus/minus stainless electrodes located at opposite ends of the horizontal tube were electrified to the point at which oxygen plasma was created. Pure helium was added as a second fill gas to $4 + 2(1.08) = 6.16$ torr so as to maintain the 4 to 1 ratio between oxygen and helium, determined by the respective mass numbers (O-16/He-4) of the gases.

Experiment 2: Helium plus Oxygen

1 part He + 4 parts O

Pressure O_2 = 2x He

In the second experiment, which is the reverse of the above, the QR vacuum tube was filled with pure helium, isolated, and pumped down to approximately 2 torr. The stainless electrodes were electrified so as to spark and maintain helium plasma. Pure oxygen was added as a second fill gas to $2 + 4(1.08) = 6.32$ torr, again, to maintain the 1 part He to 4 part O ratio determined by their respective mass numbers.

RESULTS

RGA analysis at the beginning of each test showed the presence of the first fill gases, oxygen in Experiment 1 and helium in Experiment 2. The second fill gases, helium in Experiment 1 and oxygen in Experiment 2, were also noted. In both experiments, values corresponding to the mass assignment of argon (Source: Mass Assignment Table from MKS Instruments, manufacturer of the RGA) appeared on the RGA readout following admission of the second fill gas. No argon was used in either test. These values are shown below:

Mass	Ion	Parent Substance
18	Ar ++	argon-36
20	Ar ++	argon-40
36	Ar +	argon-36
40	Ar +	argon-40

CONCLUSION

Although preliminary, these results suggest the possibility of a low energy nuclear reaction in which one atom of helium fuses with a molecule of oxygen to produce argon:

$$_2(^{16/18}_{8}O) + {}^4_2He \rightarrow {}^{36/40}_{18}Ar$$

$$^4_2He + {}_2(^{16/18}_{8}O) \rightarrow {}^{36/40}_{18}Ar$$

Placing a target gas under vacuum, charging it with electricity, then admitting a catalyst gas may cause the mutually repulsive force existing between their respective nuclei to become instantaneously neutralized. This sudden breach may allow the centripetal Casimir force generated by the vacuum/ether to cause the nuclei to fuse and form a heavier atom. In this case two lighter atoms, oxygen and helium fuse to produce the heavier argon atom. These results are of course preliminary. Further research is needed to establish whether low energy nuclear reactions are in fact produced by the above experiments.

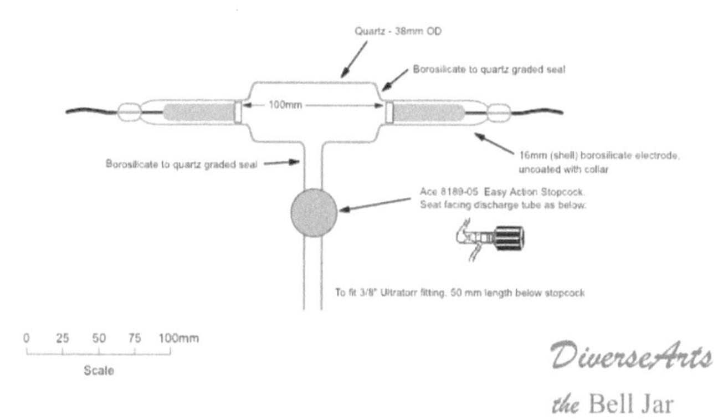

Design of vacuum tube used in Quantum Rabbit (QR) gas experiments

Helium plasma in QR tube

Source: Edward Esko, "Appearance of Argon in Oxygen/Helium Plasma," *Infinite Energy* Issue 81, 2008.

3

APPEARANCE OF COPPER ON A STAINLESS ELECTRODE

ABSTRACT

In a study conducted at Quantum Rabbit (QR) lab in Nashua, New Hampshire, USA, on May 30, 2008, QR researchers noted the anomalous appearance of what seemed to be copper on a stainless steel anode used in a lithium/oxygen vacuum test. The anode was sent to an outside lab, analyzed by ICP (Inductively Coupled Plasma Atomic Emission Spectroscopy), and found to contain trace amounts of copper.

BACKGROUND

In lithium studies conducted on February 29, 2008 and May 2, 2008 at the QR lab in Nashua a small piece of lithium was cut from a lithium rod 12.7 mm in diameter and approximately 165 mm long and placed in a glass vacuum tube. The lithium rods were from Alfa Aesar (Stock 10773/Lot E31S039). The Certificate of Analysis (C of A) of the lithium rods showed the following values:

Composition of Lithium Rod*

Lithium	100% (minus the trace elements below)
Sodium	31
Calcium	17

Iron	1
Silicon	4
Chlorine	16
Nitrogen	102
Potassium	36

*Values given in parts per million (ppm) unless otherwise noted.

The tube was pumped down to vacuum and backfilled with pure oxygen to about 3 torr. Stainless steel electrodes at either end of the tube were electrified by batteries charged by a HUBERT® portable solar generator. The samples were heated from the outside by a hand torch, causing them to vaporize. Lithium samples were retrieved, stored in mineral oil and sent to New Hampshire Materials Laboratory (NHML) for EDS and ICP analysis. The February 29th lithium test sample (NHML File 24901) was found to contain sodium at 0.94% and potassium at 0.14%, a substantial increase over the values in the C of A. ICP analysis of the May 2 lithium test samples showed similar results (NHML File 25125): *

Element	**Control**	**Test 1**	**Test 2**
Sodium	50	213	8900
Potassium	5	5	6

*Parts per million

Upon review, QR researchers determined that the increase in sodium was due to one or both of two factors:

1. A low energy nuclear reaction between lithium and oxygen producing the heavier element sodium:

$$^{7}Li + {}^{16}O \rightarrow {}^{23}Na$$
Lithium-7 + oxygen-16 → sodium-23

2. The release of sodium contained in the Pyrex (borosilicate) glass used in the tube. Wayne Martin, of M & M Glassblowing in Nashua, fabricated the glass tubes used in these studies. Wayne pointed out that the tubes were composed primarily of SiO_2 and B_2O_3. The glass also contained anywhere from 2% to 3% Al_2O_3, and most significant to our results, sodium and potassium oxides (Na_2O K_2O) at up to 4.2%. Recent studies have shown that when heated, as the QR tubes were, borosilicate glass would release sodium.

Another series of lithium tests was set for May 30, with the goal being to control for sodium. A new tube was designed so that the 60 mm center section would be made of quartz. The type of quartz used in the new tube contained a very small percentage of sodium (0.7 ppm), much less than that found in borosilicate. We felt comfortable that this trace amount wouldn't have a noticeable effect on the outcome of the test.

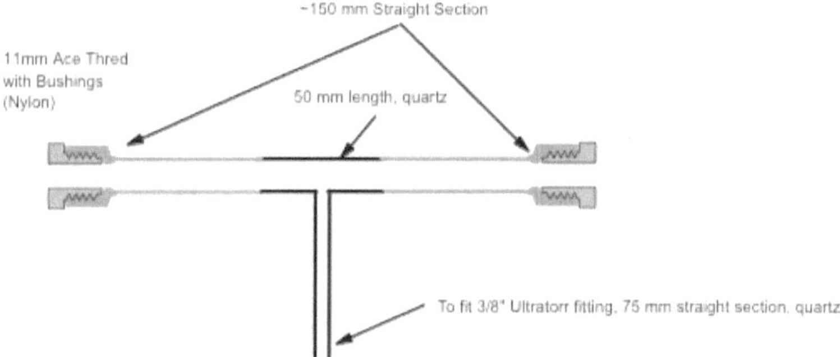

Tube design used in QR metal vapor studies. Stainless electrodes were placed at either end and test material placed in the quartz midsection between anode and cathode

METHOD

A small piece of lithium was cut from a fresh lithium rod and placed in the middle of the quartz section of the QR tube (see illustration.) Stainless electrodes (anode and cathode) were inserted at each end of the tube so that the tip of the anode came into contact with the lithium pellet. The tube was pumped down to vacuum and backfilled with pure oxygen to approximately 3.5 torr. A charge was passed through the electrodes producing a glow discharge. The sample was heated from the outside by a hand torch, at which point the metal appeared to vaporize. After several minutes, the electricity was shut off, the tube returned to atmosphere, and the sample allowed to cool.

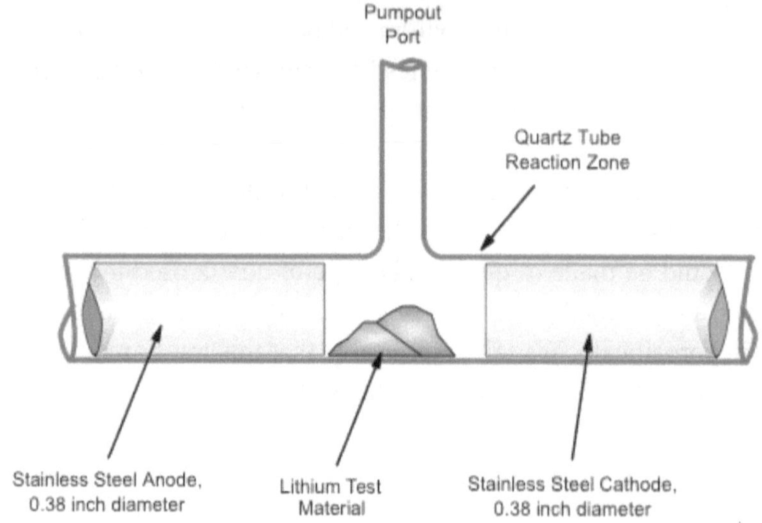

Close-up of the stainless electrodes and lithium sample used on May 30

RESULTS

The anode, cathode, and lithium pellet were removed from the tube and visually inspected. At the tip of the anode in the region of the anode that had been in contact with the lithium pellet, there appeared a copper-colored residue. This was noted and the samples packaged and sent to New Hampshire Materials Laboratory for EDS and ICP analysis. ICP analysis showed no detectable presence of sodium. This result added credence to the theory that the sodium detected in the earlier lithium tests had come from the borosilicate glass rather than from a low energy nuclear reaction between lithium and oxygen. However, consistent with the visual examination, the anode sample was found to contain 1500 ppm copper (NHML File 25237). As revealed in the C of A, the test sample of lithium contained no copper. The stainless anode also did not contain copper. The electrodes were composed primarily of iron. Their composition is as follows (shown in percentages):

Carbon	0.15 max
Chromium	17 - 19
Iron	Balance
Manganese	2 max
Nickel	8 - 10
Phosphorus	0.045 max
Silicon	1 max
Sulfur	0.03 max

Source: Diverse Arts/The Bell Jar

The quartz used in the tube contained a very small amount of copper—0.05 ppm—in addition to other trace elements in the ppm range. However, the amount of copper found on the electrode—1500 ppm—was *thirty thousand* times that amount. The experiment apparently produced an unexpected, dramatic, and unexplained increase in the element copper.

CONCLUSION

This result suggests the possibility of a low energy nuclear reaction between the pure lithium pellet and the iron in the stainless anode. During the experiment, atoms of lithium may have reacted with atoms of iron to produce atoms of copper:

$$^{7}\text{Li} + ^{56}\text{Fe} \rightarrow ^{63}\text{Cu}$$
Lithium-7 + iron-56 → copper-63

Placing the lithium in contact with the iron-rich anode, pumping down to vacuum, admitting a catalyst gas (in this case oxygen), and electrifying the tube may have caused the mutually repulsive force existing between the lithium and iron nuclei to become instantaneously neutralized. This sudden breach may have allowed the centripetal Casimir force generated by the vacuum/ether to cause the nuclei to react and form a heavier atom. In this case, two lighter atoms—lithium and iron—may have reacted to produce a heavier atom of copper. Although preliminary, this result justifies further research to determine whether or not a low energy nuclear reaction is indeed taking place in this experiment.

Afterword—Verification

The QR research team repeated the experiment on Dec. 30, 2008. The results confirmed those of May 30. Copper appeared on the stainless anode at 262 ppm and on the stainless cathode at 315 ppm. New Hampshire Materials Lab, File 25985, conducted analysis of the stainless electrodes.

Source: Edward Esko, "Appearance of Copper on a Stainless Electrode," *Infinite Energy* Issue 86, July/August 2009.

4

Appearance of Palladium on a Zinc Anode

Abstract

In a study conducted at Quantum Rabbit (QR) lab in Nashua, New Hampshire, USA, on September 30, 2008, the QR research team, myself, together with Alex Jack and Woody Johnson conducted a vacuum discharge test with a zinc anode, copper cathode, and pure sulfur. Oxygen was added as a fill gas. Upon analysis by ICP (Inductively Coupled Plasma Atomic Emission Spectroscopy) at New Hampshire Materials Laboratory, test samples were found to contain significant traces of palladium.

Background

Earlier studies at QR labs have suggested the possibility of inducing low energy nuclear reactions under relatively low temperatures and pressures, and with relatively small inputs of energy. In a previous study, QR researchers noted what appeared to be copper residue on the stainless electrode used in a vacuum test. Upon ICP analysis, the stainless anode was found to contain copper at 1500 ppm. No copper was used in the test. The result suggested the possibility of a low energy nuclear reaction, in which iron in the stainless electrode reacted with the lithium pellet used in the test. With the copper experiment as background, I formulated a new experiment designed to test the hypothesis that low energy nuclear reactions are taking place in QR vacuum discharge tests. I predicted that, under the right conditions, similar to those in the previous appearance of copper test, an atom of zinc could be made to react with an atom of sulfur to form palladium. An experiment designed to test this hypothesis was scheduled for September 30, 2008 at the QR lab in Nashua.

In a September 16, 2008 email to K.J. Thomas at the renowned Cavendish lab in England, I expressed my hope for the experiment as well as my desire to participate in cooperative research on low energy transmutation: "We would very much like to discuss the possibility of conducting a series of studies at your labs at Cavendish. We feel your facilities would allow for more precise before and after analysis, more careful control of test samples, and could extend the research to the next level, that being to detect whether the isotope distribution in test materials matches that found in nature or is unique and whether or not the low energy transmutation of elements produces energy in the form of charged particles.

"Of the various lines we have pursued, the possibility that electrode composition can be changed under vacuum and with particular catalysts is perhaps the most intriguing We have scheduled additional tests in this series at our New Hampshire lab. We plan to test zinc electrodes under vacuum in combination with sulfur and with oxygen as the fill gas. Our goal is to see whether the zinc electrodes and sulfur and oxygen catalysts can be prompted to form palladium."

METHOD

In the Sept. 30 test, we utilized the tube design employed in the lithium-iron-copper test described above. The borosilicate glass tube was 150 mm long, with a 50 mm quartz midsection. A 3/8-inch diameter quartz straight section, perpendicular to the tube and 75 mm in length, connected the midsection with the vacuum manifold, and served as the entrance for a pure oxygen backfill. We had ordered a high purity copper rod from Alfa Aesar and cut the rod to create the anode and cathode. This Puratronic® rod is composed of 99.999% pure copper (metals basis.) We also ordered a high purity zinc rod (99.9999% metals basis), and used the rod to create a pure zinc insert pressed into a 17/64-inch diameter cavity in the center of the copper anode.

First, we inserted the copper cathode in one end of the tube. Pure sulfur was placed in the quartz midsection. The sulfur, listed in the Alfa Aesar catalogue as "sulfur pieces," was actually a coarse yellow powder that was difficult to maneuver into the tube. However, after persistent attempts, our vacuum consultant succeeded in placing a sufficient quantity in the tube.

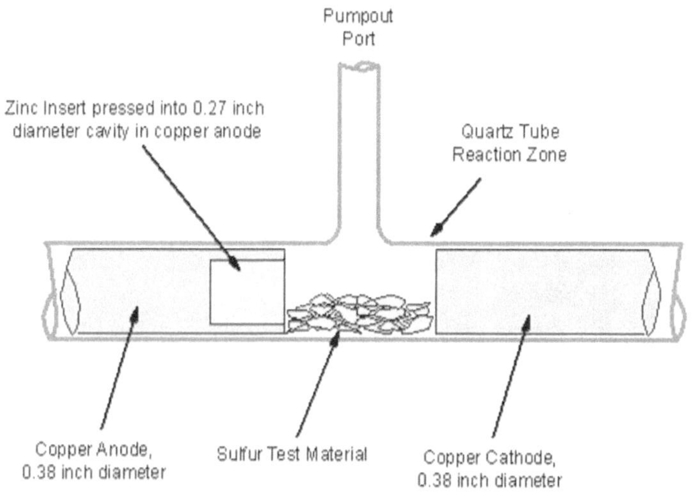

Quantum Rabbit tube and electrode configuration used in the Sept. 30 test

Like the copper electrodes, the sulfur was labeled Puratronic® and certified by Alfa Aesar as being 99.999% pure (metals basis.) After placing the sulfur in the tube, the zinc-tipped anode was inserted at the opposite end, and the electrodes connected to the power supply. Both anode and cathode were in contact with the sulfur powder. The tube was pumped down to vacuum and backfilled with oxygen to approximately 3.5 torr. Electricity was passed through the electrodes, producing an electric arc and glow discharge. The glow had a distinctive blue-white hue. A portion of the sulfur had apparently vaporized. A digital camera recorded the sulfur glow. After several minutes, the electricity was turned off, the vacuum pumps disconnected, and the sample allowed to cool.

Palladium Results

Three samples were retrieved from the tube: the zinc-tipped anode, the copper cathode, and a blackened residue from the center of the tube. The anode and cathode tips had undergone a noticeable change during the experiment, with the presence of an undetermined residue on their surface. All three samples were packaged and sent to New Hampshire Materials Laboratory for ICP analysis.

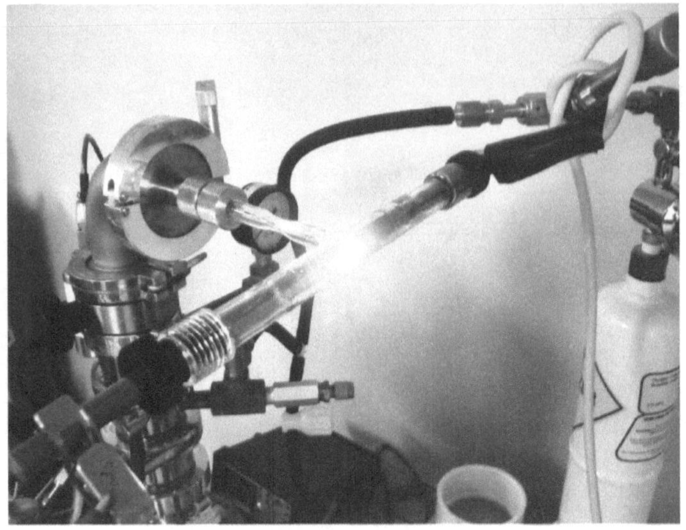

Sulfur discharge glow in QR tube Sept. 30 2008

ICP results came back a month later (NHML File 25670a dated October 30, 2008). The results appeared to confirm my prediction about the appearance of palladium following a possible low energy nuclear reaction between zinc and sulfur. Here are the differences in the amount of palladium detected in the test samples before and after the experiment. The palladium (Pd) values shown in the "Before" column are from the Certificate of Analysis (C of A) provided by Alfa Aesar for each sample. They reflect the composition of the samples prior to the test. The Pd values in the "After" column are from the NHML ICP analysis report cited above and reflect composition values following the test.

	Pd Before	**Pd After**
Zinc anode	ND	50 ppm*
Copper cathode	<0.002 ppm	40 ppm
Sulfur residue	ND	91 ppm

*Parts per million

According to the Certificates of Analysis, there was no trace (ND) of palladium in the zinc anode (Stock 12718/Lot LI4R006) and sulfur powder (Stock 10755/Lot D03S015). The amount detected in the copper cathode was cited at 0.002 ppm (Stock 10156/Lot L08R011). After the test, the amount of palladium on the copper cathode jumped to 40 ppm. Moreover, the experiment apparently led to a sudden and dramatic appearance of palladium on the zinc anode tip and in the residue; from zero to 50 ppm on the anode, and zero to 91 ppm in the residue.

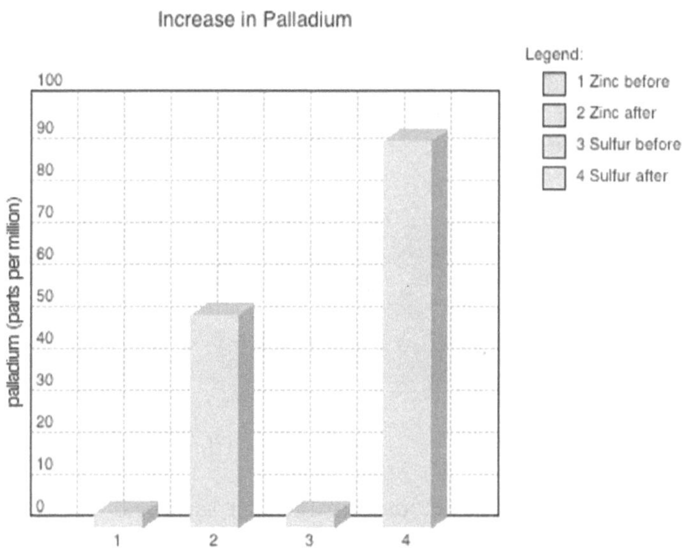

Level of palladium on the zinc anode before (1) and after (2) the Sept. 30 test, and in the sulfur residue before (3) and (4) after the test

CONCLUSION

These results may confirm the predictive power of the quantum conversion theory of low energy transmutation. As predicted, during the experiment, atoms of zinc may have reacted with sulfur to form palladium (elements shown with atomic numbers):

$$_{30}Zn + {}_{16}S \text{ (with oxygen catalyst)} \rightarrow {}_{46}Pd$$

Placing sulfur in contact with the zinc anode, pumping down to vacuum, admitting oxygen as a catalyst, and electrifying the tube may have caused the electrical repulsive force existing between two plus charged nuclei—zinc and sulfur—to become temporarily neutralized. Such temporary neutralization may have allowed the omnipresent centripetal Casimir force to cause zinc and sulfur to react and form of a heavier atom of palladium. As we see in the chart below, these three elements, zinc, sulfur, and palladium, have numerous isotopes. It would be most productive to investigate whether the isotope distribution of the palladium detected in the test matches that of palladium found in nature. Hopefully, this can become the focus of future research.

Isotope Matchups

$${}^{32}S + {}^{70}Zn \rightarrow {}^{102}Pd$$
$${}^{34}S + {}^{68}Zn \rightarrow {}^{102}Pd$$
$${}^{34}S + {}^{70}Zn \rightarrow {}^{104}Pd$$
$${}^{36}S + {}^{66}Zn \rightarrow {}^{102}Pd$$
$${}^{36}S + {}^{68}Zn \rightarrow {}^{104}Pd$$
$${}^{36}S + {}^{70}Zn \rightarrow {}^{106}Pd$$

There were other anomalies detected in the test samples. Here is the full ICP analysis from NHML.

Zinc Anode

Used in Test

Copper	1.27%
Sulfur	1.75%
Zinc	1 ppm

Detected in Treated Sample

Strontium	0.5 ppm
Palladium	50 ppm
Chromium	8 ppm

Copper Cathode

Used in Test

Copper	64.2%
Sulfur	117 ppm
Zinc	2 ppm

Detected in Treated Sample

Strontium	2 ppm
Palladium	40 ppm
Chromium	2 ppm

Residue

Used in Test

Copper	35.7%
Sulfur	7%
Zinc	3 ppm

Copper Cathode

Detected in Treated Sample

Strontium	1 ppm
Palladium	91 ppm
Chromium	<1 ppm

In addition to palladium, note the consistent presence of strontium and chromium in the test samples. According to the C of A, strontium was not present in the zinc anode or sulfur powder prior to the test. Only a minute amount—0.0001 ppm—is listed in the C of A for the copper rod. That is a vastly smaller quantity than the 2 ppm found on the copper cathode after the test. In the case of chromium, none is listed in the C of A for the zinc rod. The C of A for the sulfur powder shows chromium at less that 0.5 ppm. The C of A for the copper rod lists chromium at 0.002 ppm. Both quantities are far less than the 8 ppm found on the zinc anode after the test. The anomalous appearance of these trace metals in the test samples may be another example of the predictive power of the quantum conversion theory. Prior to the experiment, I predicted that, through the process of low energy transmutation, zinc could be converted to strontium:

$$^{70}Zn + {}^{16}O \rightarrow {}^{86}Sr$$
Zinc-70 + oxygen 16 → strontium-86

I also predicted that an atom of sulfur could be induced to react with an atom of oxygen to form an atom of chromium:

$$^{34}S + {}^{16}O \rightarrow {}^{50}Cr$$
Sulfur-34 + oxygen-16 (fill gas) → chromium-50

Having worked out these formulas in advance, I asked Tim Kenney, director of laboratory services at NHML, to scan for strontium and chromium, in addition to palladium, in the ICP analysis. Their presence in the test samples was therefore more confirmation than surprise. In another test done with a zinc anode, copper cathode, sulfur test material, and oxygen backfill conducted at QR lab on December 30, 2008, strontium appeared at 14 ppm on the zinc anode, 4 ppm on the copper cathode, and 1 ppm in the sulfur residue. Chromium appeared at 130 ppm on the zinc anode and 198 ppm in the sulfur residue (NHML File 25985.)

When I addressed the special workshop on transmutation organized by George Miley following ICCF-14 (International Conference on Cold Fusion) in August 2008 in Washington DC, I challenged the audience to conduct experiments similar to the Quantum Rabbit experiments in their laboratories. I wish to repeat that challenge. Although preliminary, the results reported here, and in my previous papers published in *Infinite Energy*, are highly suggestive. My hope is that researchers around the globe will initiate independent investigations with the potential to replicate and add credence to these original formulas and novel results. Feel free to contact me if you would like our theoretical and technical support for your research.

Source: Edward Esko, "Appearance of Palladium on a Zinc Anode," *Infinite Energy* Issue 87, September/October 2009.

5

APPEARANCE OF TIN ON A SILVER ANODE

ABSTRACT

ICP (Inductively Coupled Plasma Atomic Emission Spectroscopy) analysis of a pure silver anode used in a vacuum study conducted at Quantum Rabbit (QR) lab in Nashua, New Hampshire, USA, on September 30, 2008, revealed the anomalous appearance of tin.

BACKGROUND

Following a vacuum discharge study conducted at the QR lab on May 30, 2008, in which copper was found to have appeared on the tip of a stainless anode, I formulated an experiment designed to test the hypothesis that an atom of silver could be induced to react with an atom of lithium to form an atom of tin. The experiment was conducted on the same day (September 30) as the experiment in which palladium was found to have appeared on a zinc anode.

The silver-tin experiment was conducted by QR researchers Alex Jack, Woody Johnson, and me, with the assistance of our vacuum consultant, Steve Hansen. Our master glassblower, Wayne Martin, of M & M Glassblowing in Nashua, had fabricated the vacuum tube according to my design. Prior to the experiment, I had sent an email to K.J. Thomas at the Cavendish Lab in the UK. In that September 16 email, I described the experiment and stated our primary goal: "In the test, we plan to see whether silver and lithium can be prompted to produce tin."

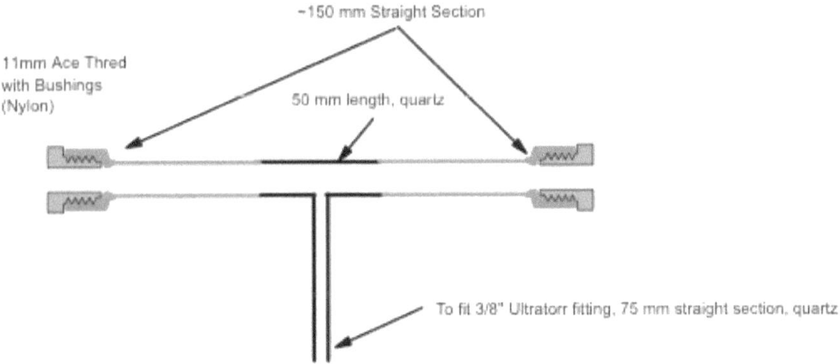

Quantum Rabbit vacuum tube used in the Sept. 30 test

METHOD

For this test, we employed the tube design used in the metal vacuum tests described above. The borosilicate glass tube was 150 mm long, with a 50 mm quartz midsection. A 3/8-inch diameter quartz strait section, perpendicular to the tube and 75 mm in length, sealed the connection to the vacuum manifold, and served as the entryway for the pure oxygen employed as a fill gas. The anode and cathode were made of a pure copper rod from Alfa Aesar. The rod is listed in the catalogue as Puratronic® (Stock 10156/Lot L08R011) and is composed of 99.999% pure copper (metals basis). There are 62 trace elements listed in the C of A (certificate of analysis) for the copper rod, most in fractions of a part per million. The list is too long to present here. Readers who are interested in obtaining the C of A for the copper rod can contact Alfa Aesar directly.

A pure silver rod was also ordered from Alfa Aesar (Stock 11473/Lot E14M22). The high purity silver rod is listed as Premion and certified at 99.999% pure. The analysis of the rod provided by the C of A shows the following trace elements, in addition to silver (Ag):

Composition of Silver Rod*

Ag	Major
Pt	1
B	ND
Ni	ND
Zn	1
Cu	2
Pd	ND
Fe	4
Pb	ND
Au	ND
Al	ND
Mg	ND
Sn	ND

*Values are given in parts per million (ppm) unless otherwise noted. ND: Sought but not detected

Please note the value given for tin (Sn): sought but not detected. This value serves as the primary point of comparison with the values detected at the outcome of the experiment. The silver rod was used to create a silver inset—0.27 inches in diameter—at the center of the copper anode. First, the copper cathode was inserted at one end of the tube. A small piece of lithium was cut from a pure lithium rod and placed in the quartz midsection. The lithium rod was from Alfa Aesar (Stock 10773/Lot E31S039). The Certificate of Analysis (C of A) of the lithium rod showed the following values:

Composition of Lithium Rod*

Lithium	100%**
Sodium	31
Calcium	17

Iron	1
Lithium	100%**
Sodium	31
Calcium	17
Iron	1
Silicon	4
Chlorine	16
Nitrogen	102
Potassium	36

*Values given in parts per million (ppm) unless otherwise noted.
**Minus the elements shown in the analysis

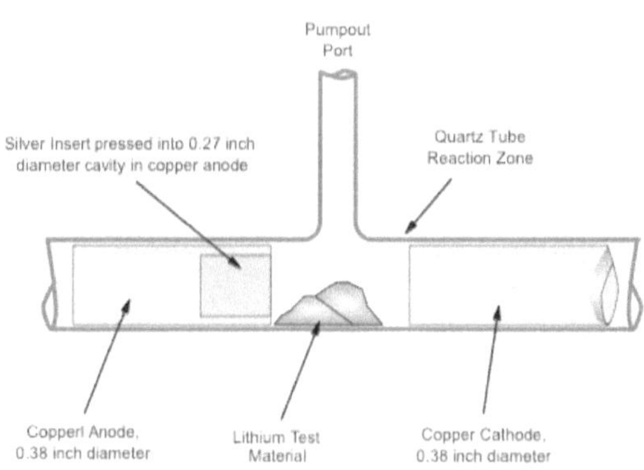

Tube and electrode configuration used in the Sept. 30 test

Following insertion of the lithium, the silver-tipped anode was inserted in the tube so that the silver anode came into contact with the lithium. The electrodes were connected to the power supply, and the tube pumped down to vacuum and backfilled with pure oxygen to approximately 3.5 torr. A charge was passed through the electrodes producing a glow discharge. The sample was heated from the outside by a hand torch, at which point the metal appeared to vaporize, giving off the ruby red color characteristic of lithium. After several minutes, the electricity was shut off, the tube returned to atmosphere, and the sample allowed to cool.

Results

The anode, cathode, and lithium pellet were removed from the tube and visually inspected. The tips of both electrodes had undergone obvious changes, as had the lithium pellet. These changes were noted and the samples packaged and sent to New Hampshire Materials Laboratory (NHML) for EDS and ICP analysis.

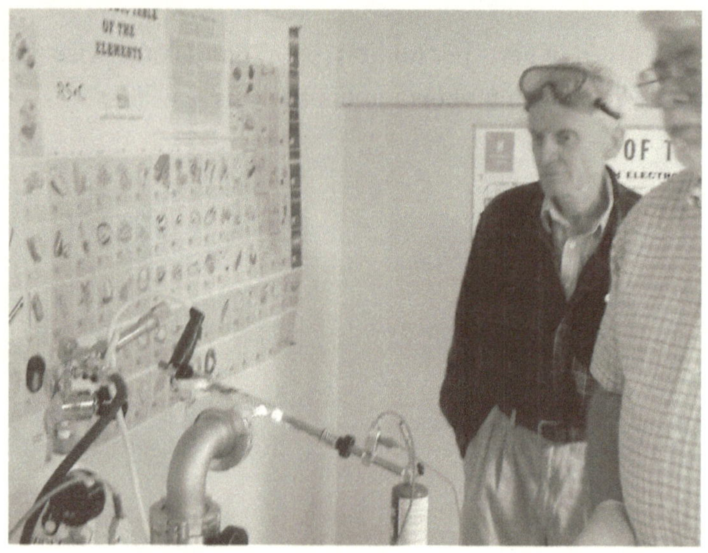

Woody Johnson (left) and Steve Hansen observe lithium
vapor in the QR vacuum tube, Sept. 30, 2008

The Test Report (NHML File 25670a) revealed the following values, as the result of chemical analysis by ICP:

<u>Analysis Results</u>

Silver Anode

Copper	65 ppm
Tin	2 ppm

Germanium   <144 ppm
Silver   38.12%

Copper Cathode
Copper   89.19%
Germanium   <38 ppm

Lithium Pellet
Copper   1050 ppm
Tin   3 ppm
Germanium   <70 ppm

CONCLUSION

As predicted by Quantum Conversion Theory, tin was found in the silver anode and lithium pellet. Although the amounts of tin detected in the treated samples are small, in the parts per million ranges, they represent a significant increase over the amounts present prior to the experiment. If we refer to the analyses of the silver and lithium rods used in the test, we see that no traces of tin are present. Moreover, there are no traces of tin in the borosilicate and quartz glass used in the vacuum tube, according to analysis provided by Wayne Martin. The only possible source of tin in the experiment was the copper rod. However, the amount of tin in the copper rod (listed in the C of A) is far less than that found in the materials after the test: less than 0.01 ppm in the copper rod before the test, versus 2 ppm in the silver anode and 3 ppm in the lithium pellet after the test, a 200- to 300-fold increase. The dramatic increase in tin suggests the possibility of a low energy nuclear reaction between lithium and silver, with oxygen as the catalyst (elements shown with atomic numbers):

$$_{47}Ag + {}_{3}Li \text{ (with oxygen catalyst)} \rightarrow {}_{50}Sn$$

Placing lithium in contact with the silver anode, pumping down to vacuum, admitting a catalyst gas (oxygen), and electrifying the tube may have caused the repulsive force existing between positively charged silver and lithium nuclei to become temporarily neutralized. This temporary breach may have allowed the centripetal Casimir force generated by the vacuum/ether to cause the nuclei to react and form a heavier atom. In this case, two lighter atoms—lithium and silver—may have reacted to form a heavier atom of tin. As we see below, these three elements—silver, lithium, and tin—have isotopes that could provide pathways for low energy transmutations. Although preliminary, these results justify further research to determine whether or not a low energy nuclear reaction between silver and lithium is indeed taking place in the experiment.

Isotope Pathways

Silver	Lithium	Tin
^{109}Ag	^{6}Li	^{115}Sn
^{107}Ag	^{7}Li	^{114}Sn
^{109}Ag	^{7}Li	^{116}Sn

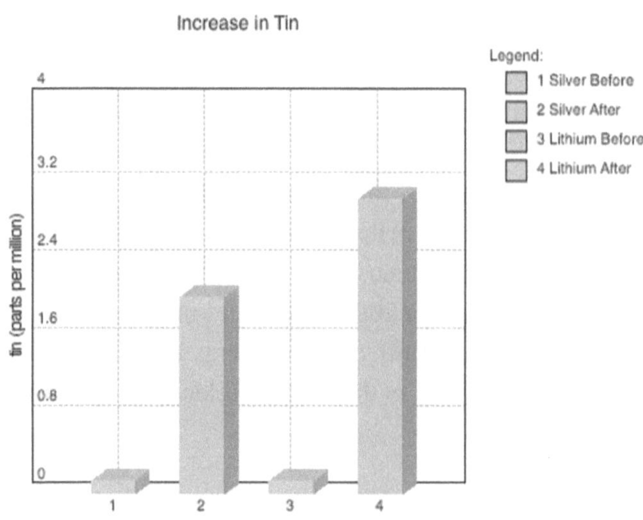

The chart shows the increases in tin in parts per million on the silver rod before (1) and after (2) the Sept. 30 test; and in the lithium metal before (3) and after (4) the test

THE GERMANIUM QUESTION

In addition to the above result, I had predicted the possibility that germanium (Ge) would be produced during the experiment. I expressed this possibility in my September 16 email to K.J. Thomas: "We plan to use a copper anode and silver cathode under vacuum, in combination with lithium and oxygen as the fill gas…to see whether copper and lithium can be prompted to produce trace amounts of germanium." The prediction was based on the Quantum Conversion formula:

$$^{63}Cu + {}^{7}Li \text{ (with oxygen catalyst)} \rightarrow {}^{70}Ge$$
$$\text{Copper-63 + lithium-7} \rightarrow \text{germanium-70}$$

It was fo-r this reason that I asked Tim Kenny at New Hampshire Materials Lab to test for germanium in the samples. However, as we see in the analysis report, the germanium result is inconclusive. The amounts indicated in the report represent the detection limits for germanium, which, in comparison to many other elements, are high. Therefore, we weren't able to say with precision whether or not germanium was present in the test samples. However, in vacuum tests run on December 30, 2008, in which copper electrodes, lithium test material, and oxygen backfill were employed, as in the Sept. 30 test, germanium appeared as follows (ICP analysis by New Hampshire Materials Lab, File 25985):

Copper Anode	Copper Cathode	Lithium Residue
260 ppm	50 ppm	106 ppm/154 ppm (in a second test)

These results suggest the possibility of the low energy nuclear reaction postulated above.

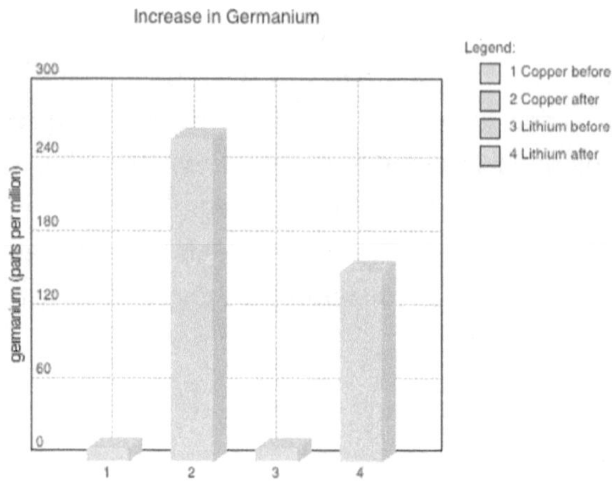

Increase in germanium in the Dec. 30, 2008 study

The chart shows the increases in germanium in parts per million on the copper anode before (1) and after (2) the Dec. 30 test; and in the lithium metal before (3) and after (4) the test. The amount in the copper prior to the test was estimated to be 0.02 ppm and 0 ppm in the lithium.

AFTERWORD—VERIFICATION

Perhaps the most dramatic germanium result occurred in a July 30, 2009 test sponsored by the New Energy Foundation and conducted at the Quantum Rabbit lab in Owls Head, Maine. Copper electrodes were used in the test, as was a lithium pellet and pure oxygen backfill. ICP conducted by NHML revealed germanium at 2190 ppm on the copper cathode and 388 ppm on the lithium pellet (NHML File 26657/August 1, 2009), adding further confirmation of a possible low energy nuclear transmutation taking place in Quantum Rabbit studies.

Source: Edward Esko, "Appearance of Tin on a Silver Anode," *Infinite Energy* Issue 88, November/December 2009.

6

CARBON ARC UNDER VACUUM

ABSTRACT/BACKGROUND

Researchers at Quantum Rabbit (QR) LLC in the United States have repeatedly seen the anomalous appearance of silicon and a variety of metals, including magnesium, aluminum, scandium, iron, cobalt, and nickel in pure graphite powder. These carbon arc studies were conducted in the open air and seem to have produced a cascade of low energy nuclear reactions in which carbon (in the graphite powder and graphite rods used in the arcing process) reacted with oxygen and nitrogen. A summary of this research—"Production of Metals from Non-Metallic Graphite"—was published *in Infinite Energy*, Issue 78, 2008. On December 30, 2008, QR researchers achieved similar results, notably the appearance of magnesium, silicon, iron and other elements on the surface of the pure graphite rods used in a carbon arc test performed under vacuum. The study took place at the QR lab in Nashua, New Hampshire, USA.

METHOD

Unlike previous carbon arc studies conducted in open air, the Dec. 30 test took place in a specially designed vacuum tube. The tube design was that utilized in previous QR vacuum studies. The borosilicate tube was 150 mm long, with a 60 mm quartz midsection. A 3/8-inch diameter quartz straight section, perpendicular to the tube and 75 mm in length, connected the quartz reaction zone with the vacuum manifold, and served as the entrance for a pure oxygen backfill. None of the elements later found on the electrodes were in the quartz material.

We had ordered high purity graphite rods from Alfa Aesar and used the rods to create the anode and cathode. The graphite electrodes were inserted into aluminum electrodes, 0.38-inch in diameter, designed to be flush with the inner wall of the tube so as to seal the tube to maintain vacuum. The aluminum electrodes were positioned outside the quartz reaction zone so as not to interfere with reactions taking place in the center of the tube.

The graphite rods were ultra high purity, 99.9995% pure (metals basis). Their C of A is as follows (Stock 14754/Lot F27S027):

Graphite electrode, counter-flat top, 3.05mm dia,, 38.10 mm long, 99.9995% (metals basis)

Al	ND	Fe	ND	Ni	ND	W	ND
B	ND	Pb	ND	Si	ND	V	ND
Ca	ND	Mg	ND	Ag	ND	Zn	ND
Cr	ND	Mn	ND	Sn	ND	Zr	ND
Ca	ND	Mo	ND	Ti	ND		

Values given in ppm unless otherwise noted
ND: Not detected

First, the graphite cathode was secured in the tube, and a catalyst of pure sulfur was placed in the quartz midsection. The sulfur, listed in the Alfa Aesar catalogue as "sulfur pieces," was actually a coarse powder. It was labeled as Puratronic® (Stock 10755/Lot D03S015) and certified as being 99.999% pure (metals basis.) After the sulfur was placed in the tube, the graphite anode was inserted in the opposite end, and the electrodes connected to the power supply. The tube was pumped down to vacuum and backfilled with pure oxygen to approximately 3.5 torr. Electricity was passed through the electrodes, producing an electric arc and glow discharge with the blue/white characteristic of sulfur. After several minutes, the electricity was turned off, the vacuum pumps disconnected, and the sample allowed to cool.

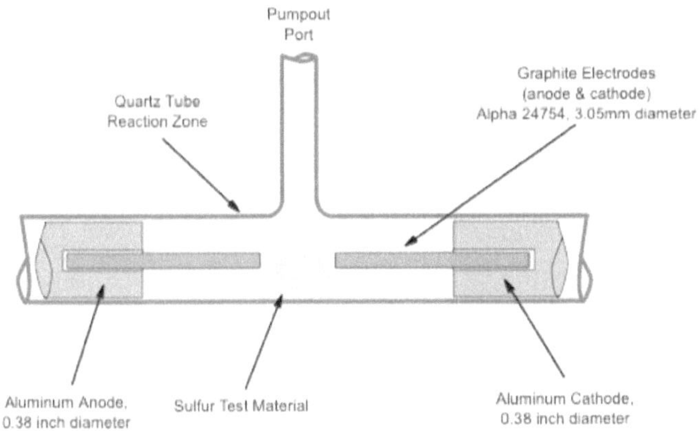

Tube and electrode configuration used in the Dec. 30 test

The graphite anode and cathode tips had undergone noticeable changes during the arcing process, including the presence of an undetermined residue on their surface. Both graphite electrodes were packaged and sent to New Hampshire Materials Laboratory for analysis by ICP (Inductively Coupled Plasma Atomic Emission Spectroscopy). ICP results came back on January 9, 2009 (NHML File #25985). Results were consistent with previous carbon arc studies and are presented below.

Magnesium

No magnesium was listed in the C of A for the graphite rods. The C of A for the sulfur pieces listed magnesium at 0.2 ppm. The treated anode came back with a magnesium content of 7 ppm; and the cathode, 8 ppm (see table "Increase in Magnesium.") The presence of magnesium was consistent with open-air studies, one of which (August 9, 2007, NHML File 24067) revealed magnesium at up to 1800 ppm. The appearance of magnesium across a wide spectrum of carbon arc studies suggests a basic nuclear reaction:

$$^{12}C + {}^{12}C \rightarrow {}^{24}Mg$$

Carbon-12 + carbon-12 $\rightarrow$ magnesium-24

Simply put, when exposed to arcing, both in open air and under vacuum, two atoms of carbon may react with each other to form an atom of magnesium.

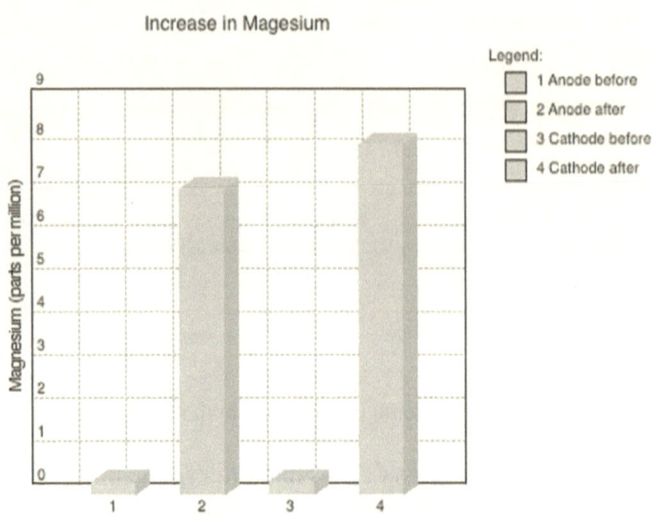

Levels of magnesium in anode and cathode before and after the test

SILICON

Together with magnesium, silicon has appeared consistently in carbon arc studies. In the open-air study cited above (NHML File 24067), silicon appeared in treated graphite at 10,500 ppm, or 1.5%. ICP of the treated anode from Dec. 30 revealed silicon at 61 ppm and of the treated cathode at 138 ppm (see table "Increase in Silicon.") No silicon was listed in the C of A for the graphite rods. Analysis of the sulfur pieces revealed silicon at less than 1 ppm.

The consistent appearance of silicon suggests a fundamental relationship existing between carbon and oxygen. As we know from chemistry, carbon and oxygen readily combine to form compounds like carbon dioxide and the organic compounds that form the basis of life. Apparently, given the proper conditions, the mutual attraction existing between these two elements can prompt them to react and form the heavier atom silicon. The formula for this transmutation is as follows:

$$^{12}C + {}^{16}O \rightarrow {}^{28}Si$$

Carbon-12 + oxygen-16 $\rightarrow$ silicon-28

Apparently, the transmutation of carbon and oxygen into silicon can be achieved under relatively low pressures, temperatures, and inputs of energy.

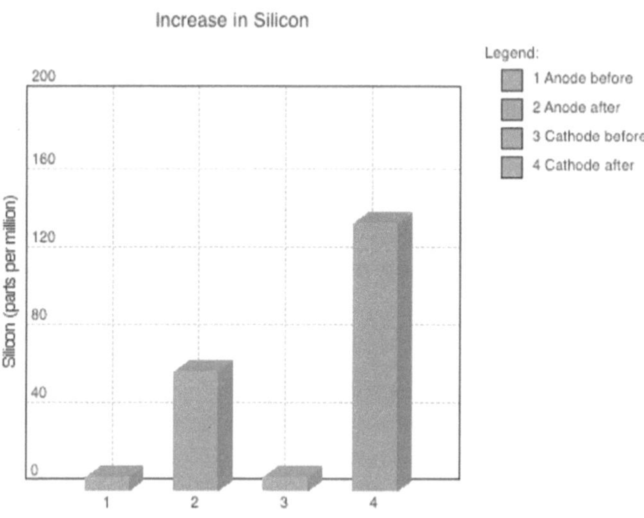

Levels of silicon in anode and cathode before and after the test

ANOTHER SILICON PATHWAY

As part of the same lab session on Dec. 30, the QR team performed another test that may hint at a different pathway to silicon. In this test, inserts of lithium and sodium were pressed into pure copper electrodes. Anode and cathode were fabricated out of a pure copper rod. Sulfur test material was placed between the anode and cathode and the tube pumped down to vacuum. Anode and cathode were electrified, producing a glow discharge. After several minutes, the power was disconnected, the tube returned to atmosphere, and the test materials allowed to cool. Prior to the test, I had predicted that silicon could be formed through the low energy reaction of lithium and sodium, according to the following formula:

^{6}Li + ^{23}Na (with oxygen catalyst) $\rightarrow$ ^{29}Si
Lithium-6 + sodium-23 $\rightarrow$ silicon-29

As expected, silicon appeared in the test samples, as did cobalt (Co), which I predicted would appear as the result of a low energy reaction between sodium and sulfur:

^{23}Na + ^{36}S (with oxygen catalyst) $\rightarrow$ ^{59}Co
Sodium-23 + sulfur-36 $\rightarrow$ cobalt-59

Cobalt could also be the product of a reaction between an atom of sodium-23 and two atoms of oxygen-18 (either from the pure oxygen fill, or through the low energy fission of one atom of sulfur into two atoms of oxygen:

^{36}S (with oxygen catalyst) $\rightarrow$ ^{18}O + ^{18}O
^{23}Na + $_2$(^{18}O) $\rightarrow$ ^{59}Co

Silicon and cobalt appeared in the test samples with the following values in parts per million (NHML File 25985):

	Anode	**Cathode**	**Sulfur Residue**
Si	190	45	340
Co	5	4	1

According to the Certificates of Analysis from Alfa Aesar, the lithium rod contained silicon at 4 ppm, the sulfur pieces <1 ppm, and the copper electrodes <0.005 ppm. Moreover, cobalt was present only in the copper electrodes but in miniscule amounts, <0.0005 ppm. These quantities are far below the quantities found in the test materials. We should note, however, that silicon is a constituent of the quartz vacuum tube. Thus, in order to more accurately control for silicon, these experiments would need to be conducted in a vacuum chamber made of material in which silicon is not a component.

IRON

The presence of iron in treated graphite may help explain the magnetic properties consistently seen in treated samples. Evidence suggests this magnetic effect may be permanent, having been observed in graphite powder more than a year after the carbon arc test. In the open-air test cited above (NHML File 24067) iron was found at 4700 ppm. It also appeared in the Dec. 30 vacuum test, even though it was not listed in the C of A for the graphite rod. (Iron was listed in the C of A for the sulfur at less than 1 ppm.) ICP analysis on the Dec. 30 graphite samples revealed iron at 8 ppm on the anode and 5 ppm on the cathode (see chart "Increase in Iron.")

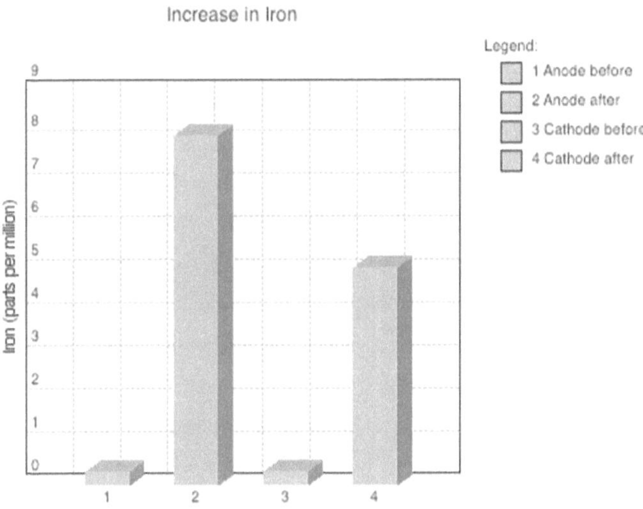

Levels of iron in anode and cathode before and after the test

The formation of iron may be an example of the phenomenon known as *proton emission*, a type of radioactive decay in which one or more protons are ejected from a proton-rich nucleus. In Quantum Conversion Theory, iron is formed when two atoms of carbon react with two atoms of oxygen:

$$2(^{12}_{6}C + ^{16}_{8}O) \rightarrow\, ^{56}_{26}Fe\; (+\,2\;protons)$$

In this reaction, the transmuted nucleus has two too many protons (28) and simultaneously ejects the two extra protons to restore equilibrium ($28 - 2 \rightarrow {}^{26}Fe$). The protons tunnel out of the nucleus in finite time, in a process known as *quantum tunneling*. To date, more than twenty-five isotopes have been found to exhibit proton emission; proton emitters are not seen in naturally occurring isotopes but are produced via nuclear reactions such as those taking place during the carbon arcing process. The simultaneous emission of two protons (double proton decay) was not observed until 2002 when it was found in an isotope of iron. Other magnetic isotopes can be produced in this reaction. Nickel has appeared consistently in carbon arc experiments, as has cobalt. Nickel was found in the treated anode at 6 ppm; cobalt in the treated cathode at 2 ppm, although neither was listed in the C of A.

It is possible that nickel (Ni-58) is formed by the reaction of two atoms of carbon-12 with two atoms of oxgyen-17, and by the reaction of two atoms of carbon-12 with two atoms of oxygen-18 (Ni-60.) Cobalt-59 may be produced when atoms of nickel-58 or nickel-60 shed one proton and one or more electrons or neutrons through quantum tunneling. More research is needed to determine the pathways leading to the appearance of these elements.

CARBON-NITROGEN REACTIONS

The vacuum test conducted on Dec. 30 was consistent with previous open-air tests in that the same reactions between carbon and oxygen were observed. However, since the test was conducted under vacuum, with pure oxygen backfill, reactions between carbon and nitrogen were non-existent. For example, aluminum appears consistently in open-air studies but was not found in the vacuum test. In the open-air test cited above (NHML File 24067), aluminum was found in the treated graphite at 7800 ppm. The Quantum Conversion formula for the low energy production of aluminum is as follows:

$${}^{12}C + {}^{15}N \rightarrow {}^{27}Al$$

Carbon-12 + nitrogen-15 → aluminum-27

Silicon (Si-30) can also react with nitrogen (N-15) to produce scandium (Sc-45), an element that has consistently appeared in open-air graphite studies. Perhaps another round of vacuum studies can be designed in which nitrogen is substituted for oxygen as the fill gas. These studies would test the hypothesis that, in addition to reacting with oxygen, carbon reacts with nitrogen to form new elements.

CONCLUSION

As stated in the introduction, the carbon arc studies conducted under vacuum yielded results consistent with those conducted in the open air, with several important differences, both in the experimental inputs and in the results, as summarized in the table.

Table: Comparison of Carbon Arc Study Methods

Open Air	Vacuum
Greater power input (36-48 Volts DC)	Smaller power input (Approx. 4 amps DC)
Higher temperatures (Graphite electrodes glow white hot)	Lower temperatures (Little or no electrode glow)
Higher pressure (1 atmosphere = 760 torr)	Lower pressure (Approx. 3.5 torr)
Graphite electrodes plus graphite powder (Greater carbon reactivity)	Graphite electrodes only (Less carbon reactivity)
O^2 and N^2 as catalyst gases	O^2 as single catalyst gas (plus solid sulfur)

Open Air	**Vacuum**
Wider range of LENR (C + O and C + N reactions)	Narrower range of LENR (C + O reactions only)
Higher volume of converted elements (Mg, Si, Fe, Co, Ni,etc.)	Lower volume of converted elements
Samples test positive for magnetic properties	Magnetic activity untested

Apparently, the stronger electric arc inputs present in the open-air tests, in combination with higher temperatures, the use of fine mesh graphite powders, and the availability of oxygen and nitrogen in the atmosphere, produced a wider array of low energy nuclear reactions with much higher rates of quantum conversion than a parallel test conducted under vacuum. More research utilizing both methods is of course needed to test this hypothesis and confirm or refute these results.

Source: Edward Esko, "Carbon Arc Under Vacuum," *Infinite Energy* Issue 90, March/April 2010.

7

Appearance of Potassium in a Li-S Matrix

Abstract/Background

In a study conducted at Quantum Rabbit (QR) lab in Nashua, New Hampshire, USA on February 29, 2008, a lithium test sample was found to contain potassium at 0.14%, a substantial increase over the 36 ppm value for potassium listed in the C of A for the pure lithium rods used to create the sample. However, this result was inconclusive due to the fact that the Pyrex (borosilicate) glass used in the vacuum tube contained sodium and potassium oxides at up to 4.2% (Source: M & M Glassblowing, Nashua, NH.) QR researchers were attempting to prove the low energy nuclear reaction: $^7Li + _2(^{16}O) \rightarrow {}^{39}K$ (lithium-7 reacts with two atoms of oxygen-16 to form an atom of potassium-39.)

In subsequent studies conducted at the Nashua lab and at the QR lab in Owls Head, Maine, on December 30, 2008 and July 30, 2009 respectively, a significant trace of potassium was found on copper electrodes and in lithium/sulfur test material. The amount of potassium detected by ICP (Inductively Coupled Plasma Atomic Spectroscopy) was significantly greater than the amount in the copper electrodes, lithium, or sulfur prior to the test, or the 0.6 ppm present in the quartz used to form the central reaction zone of the glass vacuum tubes.

In order to control for potassium, tubes with a 50 mm center section made of quartz were used. The quartz contained a minute trace of potassium, not enough we believed, to influence the outcome of the test. The quartz tube was used in both lab sessions. The tests were conducted by QR researchers Alex Jack, Woody Johnson, and me, with technical assistance from Bill Zebuhr.

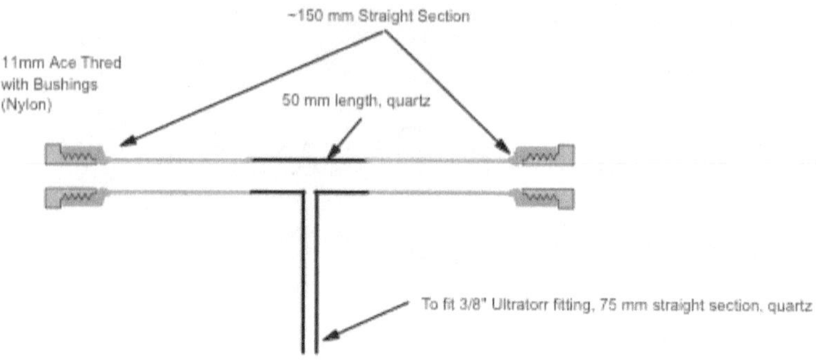

Design of the QR vacuum tube

METHOD

The same method was employed in the Dec. 30 and July 30 tests. A high purity copper cathode (from Alfa Aesar) was inserted in the QR vacuum tube. The copper was labeled Puratronic® and composed of 99.999% pure copper (metals basis). A small piece of lithium was cut from a fresh lithium rod and placed in the middle section of the tube, together with pure sulfur powder. Like the copper electrodes, the sulfur was labeled Puratronic® and certified as being 99.999% pure.

After placing the sulfur in the tube, the copper anode was inserted at the opposite end and the electrodes connected to the power supply. Both anode and cathode were in contact with the sulfur powder. The tube was pumped down to vacuum and backfilled with oxygen to about 3.5 torr. Electricity was passed through the electrodes, producing a glow discharge. A portion of the lithium and sulfur had apparently vaporized. After several minutes, the electricity was turned off, the vacuum pumps disconnected, and the samples allowed to cool.

RESULTS

In both tests, the electrodes were removed from the tube and packaged for shipping. The tube contained a blackened residue and was also sent to the outside lab, New Hampshire Materials Laboratory, for ICP analysis. The

ICP results for the Dec. 30 test came back on January 9, 2009 (NHML File 25985). The results for July 30 came back on August 14, 2009 (NHML File 26657). Both appeared to confirm our prediction about the appearance of potassium following a low energy nuclear reaction between lithium and sulfur. Here are the differences in the amount of potassium (K) detected in the test materials before and after the experiment. All values are in parts per million (ppm).

	K Before **(C of A)**	**K After** **Dec. 30**	**K After** **July 30**
Copper anode	<0.005*	230	638
Copper cathode	<0.005*	73	85
Sulfur	None listed**	38	27
Lithium	36***	See above values for S; noted as "S-Li Residue" in analysis report	

*Alfa Aesar Stock 10156/Lot L08R011
**Alfa Aesar Stock 10755/Lot D03S015
***Alfa Aesar Stock 10773/Lot E31S039

When we add the amount of potassium found in the test materials prior to the experiment, we obtain the following total:

Copper electrodes (2)	<0.010*
Sulfur	0
Lithium	36
Quartz tube material	0.6

Total K	**<36.610**

*Parts per million

According to the Certificates of Analysis, each copper electrode contained less than 0.005 ppm potassium. There was no potassium listed in the C of A for the sulfur pieces, while the lithium rods contained 36 ppm. When added to the 0.6 ppm potassium contained in the quartz tube, the total equals less than 36.610 ppm potassium. The amount in the anode after the Dec. 30 test equaled 230 ppm, a greater than six-fold increase. On July 30, the amount of potassium on the anode jumped to 638 ppm, a more than seventeen-fold increase over the total amount in the various materials before the test. The increase is greater if we consider the total amount of potassium found in all the samples, including cathode and residue, after the test.

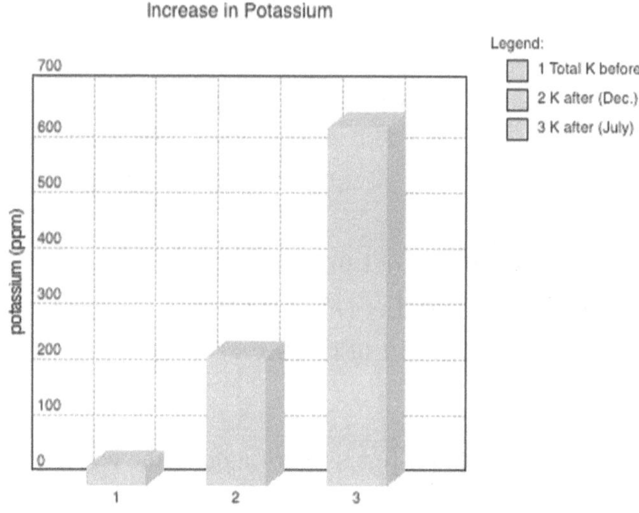

Increase in potassium (as measured on the copper anode)
in the Dec. 30 test (2) and the July 30 test (3)

CONCLUSION

The potassium results of Dec. 30, 2008 and July 30, 2009 have helped confirm the predictive power of the quantum conversion theory of low energy transmutation. As predicted prior to both tests, during the experiment, atoms of lithium may have reacted with atoms of sulfur to form potassium:

$$^7\text{Li} + {}^{32}\text{S} \text{ (with oxygen catalyst)} \rightarrow {}^{39}\text{K}$$
$$\text{Lithium-7} + \text{sulfur-32} \rightarrow \text{potassium-39}$$

Placing lithium in contact with sulfur, pumping down to vacuum, admitting oxygen as a fill gas, and electrifying the tube may have caused the electrical repulsive force between the lithium and sulfur nuclei to become temporarily neutralized. This, in turn, may have allowed the omnipresent centripetal Casimir force to cause a lithium-7 nucleus to react with a sulfur-32 nucleus to form the heavier nucleus of potassium-39. Moreover, this result may confirm the validity of the earlier formula and attempt to create potassium through a low energy nuclear reaction between lithium and oxygen. Sulfur-32 may well be formed by the low energy nuclear reaction between two atoms of oxygen-16, or $^{16}\text{O} + {}^{16}\text{O} \rightarrow {}^{32}\text{S}$. In this view, sulfur is a crystallized form of two atoms of oxygen. Given the proper conditions, it may be possible to form potassium from the reaction of an atom of lithium with two atoms of oxygen. We extend our thanks to the New Energy Foundation (NEF) whose sponsorship and technical assistance made the July 30 test at Owls Head possible.

Source: Edward Esko, "Appearance of Potassium in a Li-S Matrix," *Infinite Energy* Issue 91, May/June 2010.

8

Anomalous Metals in Electrified Vacuum

Abstract

In studies funded by the New Energy Foundation (NEF) and conducted on July 30, 2009 at Quantum Rabbit (QR) lab in Owls Head, Maine, USA, researchers performed several vacuum discharge tests utilizing a copper anode—into which a pure lead insert had been pressed—a copper cathode, pure lithium test material, and a pure sulfur catalyst. Pure oxygen was added to the vacuum tube as a fill gas. Upon analysis by an independent lab, test samples were found to include the anomalous presence of germanium (Ge) at up to 3196 ppm; potassium (K) at up to 750 ppm; and gold (Au) at up to 174 ppm. Although contamination cannot be definitively ruled out as the source of these anomalies, the possibility of low energy nuclear reactions is also a factor to be considered.

Background

Although not accepted by mainstream science, investigators around the globe have reported on the possibility of low energy nuclear reactions resulting in apparent element transmutations. These findings have been reported at international conferences, in books, in periodicals such as *Infinite Energy*, and online. In vacuum and carbon arc studies conducted between 2005 and 2009 by Quantum Rabbit LLC and published in *Infinite Energy*, independent analysis of test samples has documented the anomalous appearance of magnesium, silicon, chromium, iron, cobalt, nickel, copper, germanium, strontium, palladium, and tin on electrodes and in test materials. Although contamination of test samples cannot be ruled out, these studies are nevertheless suggestive of low energy nuclear reactions.

The Vacuum Tube

The July 30 experiments utilized the special vacuum tube designed by me and employed in previous metal vapor tests. The borosilicate glass tube was 150 mm long, with a 50 mm quartz midsection. A 3/8-inch diameter quartz straight section, perpendicular to the tube and 75 mm in length, connected the midsection with the vacuum manifold, and served as the entrance for a pure oxygen backfill.

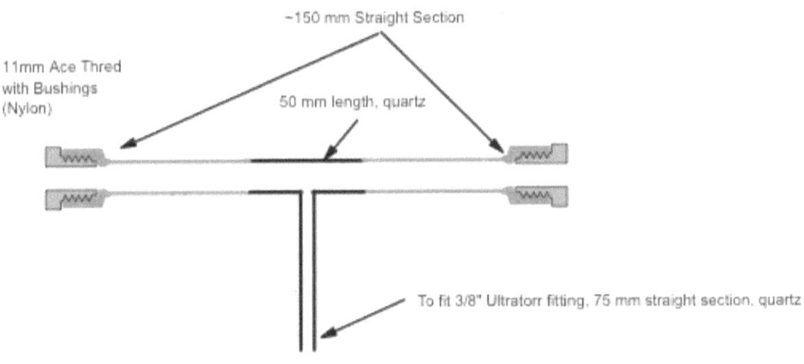

Vacuum tube used in the July 30 test

Wayne Martin of M & M Glassblowing in Nashua, NH fabricated the vacuum tubes with design assistance from Steve Hansen, founder of the Bell Jar and Diverse Arts in Owls Head, ME. The typical trace element composition of the quartz midsection of the tube, the section in which the main reaction took place, is presented in the book, *Cool Fusion*.

Copper Electrodes

A high purity copper (Cu) rod from Alfa Aesar was cut into two and used to create the anode and cathode. The Alfa Aesar catalogue stated that the Puratronic® (high purity research chemicals and materials) rod was composed of 99.999% pure copper (Stock number: 10156/Lot number:

LO8R011). The Certificate of Analysis (C of A), certified by Alfa Aesar Quality Control, is presented in the book *Cool Fusion.*

LEAD INSERT

We also ordered a high purity lead (Pb) slug (99.9999% metals basis), and used the lead slug to create a lead insert pressed into a 0.27-inch diameter cavity in the center of the copper anode. The C of A provided by the supplier is reprinted in *Cool Fusion* (Stock number: 43415/Lot number: H18T011). **Note:** In handling the lead material, all of the precautionary safety and disposal guidelines set forth in the Materials Safety Data Sheet (MSDS) provided by the supplier were observed.

LITHIUM RODS

The lithium used in the tests was cut from a pure lithium rod ordered from Alfa Aesar. The rod was listed as 99.9% pure. The C of A from the supplier is reprinted in *Cool Fusion* (Stock number: 10773/Lot number: E31S039).

HIGH PURITY SULFUR

High purity sulfur, in the form of sulfur pieces, was also ordered from Alfa Aesar and used in the test. The C of A for the sulfur, published in *Cool Fusion*, lists the sulfur as Puratronic® 99.999% pure (Stock number: 10755/Lot number: D03S015).

THE EXPERIMENT

I designed the experiment as follows: first, the copper cathode was inserted in one end of the tube. Small granules of sulfur (S) were placed in the quartz midsection, together with pieces of lithium (Li) cut from the lithium rod. The sulfur, listed in the Alfa Aesar catalogue as "sulfur pieces," was actually a coarse yellow powder that was difficult to maneuver into the tube. However, after persistent attempts, we succeeded in placing a sufficient quantity in the tube. Like the copper electrodes, the sulfur certified by Alfa Aesar as being 99.999% pure (metals basis.) The lithium pieces were also from Alfa Aesar and certified as 99.9% pure (metals basis).

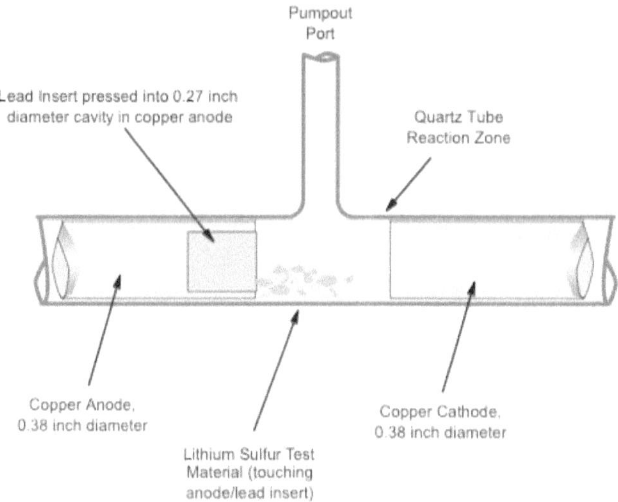

Vacuum tube and electrode configuration used in the July 30 tests.

After placing the sulfur and lithium in the tube, the lead-tipped anode was inserted at the opposite end, and the electrodes connected to the power supply. The lead-tipped anode was in contact with the lithium/sulfur powder. The tube was pumped down to vacuum and backfilled with pure oxygen to approximately 3.5 torr. Electricity was passed through the electrodes, producing an electric arc (approximately 4 amps) and glow discharge. After several minutes, the electricity was turned off, the vacuum pumps disconnected, and the sample allowed to cool.

Test Results

The experiment was conducted three times on July 30, with minor variations between experiments. Three samples were retrieved from each of the three tubes: the lead-tipped anode, the copper cathode, and sulfur-lithium residue from the center of the tube. The anode and cathode tips had undergone noticeable changes during the experiment, with the presence of an undetermined residue on their surface. The S-Li residue had also changed. All nine samples were packaged and sent to New Hampshire Materials Laboratory for ICP analysis. ICP results came back on August 14, 2009 (NHML File Number 26657). The test report is reproduced below.

New Hampshire
MATERIALS
Laboratory, Inc.

Test Report

August 14, 2009

Mr. Edward Esko   File Number: 26657
Quantum Rabbit LLC

Overview:
Samples Received: (3) Lots of residue samples as noted below
Work Requested: Chemical analysis by ICP
Sample Disposition: Return remains to client

Analysis Results:

Test 1	Lead Anode	Copper Cathode	S-Li Residue
Germanium	1018 ppm	2190 ppm	388 ppm
Potassium	638 ppm	85 ppm	27 ppm
Gold	162 ppm	11 ppm	1 ppm
Sample weight (gm)	0.0157	0.0494	0.2800

Test 2	Lead Anode	Copper Cathode	S-Li Residue
Germanium	102 ppm	119 ppm	31 ppm
Potassium	16 ppm	31 ppm	6 ppm
Gold	5 ppm	<1 ppm	<1 ppm
Sample weight (gm)	0.433	0.1047	0.7861

Test 2	Lead Anode	Copper Cathode	S-Li Residue
Germanium	35 ppm	2 ppm	7 ppm
Potassium	17 ppm	10 ppm	1 ppm
Gold	7 ppm	<1 ppm	<1 ppm
Sample weight (gm)	0.0290	0.1283	0.6961

Prepared by:
Timothy M. Kenney
Director of Laboratory Services

ANALYSIS

Where did the anomalous metals come from? According to the most common interpretation, which I refer to as the contamination theory, the anomalous elements were present in the experiment at the beginning and were not detected until the test materials were analyzed. That is a perfectly reasonable assumption given the current model of physics. It is helpful in considering this possibility to review the analysis data for the materials used in the experiment to see if and where the anomalous elements show up. (The one input for which we do not have specific data is the pure scientific grade oxygen supplied by Spec Gas, Inc. and used as backfill in the tube. Spec Gas certifies this scientific grade oxygen as 99.99999% pure. See Specgas. com for more information.)

Inputs	**Germanium**	**Potassium**	**Gold**
Quartz Tube Material	ND*	0.6 ppm	ND
Copper Electrodes	<0.02 ppm	<0.005 ppm	<0.5 ppm
Lead Insert	ND	ND	ND
Lithium Pieces	ND	36 ppm	ND
Sulfur Pieces	ND	ND	ND

*Element sought but not detected.

Outside contamination is certainly a possibility, for example, from the air in which the samples were exposed or in some manner during the handling of the materials prior to and following the tests. Also it may be that the data in the Certificates of Analysis (C of A) are unreliable, representing generic batches of product rather than the lots out of which the actual test materials were culled. For example, even though the C of A for the lead slug shows no presence of gold, certain lead alloys, such as Novodneprite ($AuPb^3$) named after the region in Kazakhstan where it is mined, occur in combination with gold. This mineral is made up of 75.94% lead and 24.06% gold. Although remote, it is possible that the actual lead sample used in the experiment came from the Novodneprovskoe Deposit and was contaminated with minute traces of gold; contamination that went undetected by the supplier but which appear in ICP analysis following the experiment.

I invite readers to review the data and suggest alternate pathways for contamination. If contamination can be proven without a doubt as the source of the anomalous metals, then we have no need for further research in this area. I challenge investigators with the correct lab apparatus and controls to prove or disprove this possibility. At the same time, ideas pointing to the possible source of contamination could lead future investigators to design carefully controlled experiments yielding more definitive results.

The other possibility can be referred to as the concentration theory. In this hypothesis, the trace amounts of the anomalous elements present in the test materials are somehow "concentrated" during the experimental process so that they appear in higher concentrations in the final analysis. So for example, potassium appears in the lithium test material at 36 ppm, the quartz tube material at 0.6 ppm, and the copper electrodes at <0.005 ppm. The electrification and heating of the test materials may concentrate the already existing potassium in the test sample, especially in the quartz used for the part of the tube that came into contact with test material. However, whether an experiment such as this is capable of concentrating potassium from the combined total of 36.605 ppm documented in the materials prior to the experiment to the 638 ppm recorded after the experiment remains to be seen.

Since germanium is listed as occurring in the copper rod at <0.02 ppm, it is difficult to see how the concentration theory would apply, unless, once again, the Certificate of Analysis provided by the supplier is unreliable. Please note that the amount of germanium detected on copper cathode following Test 1 is 2190 ppm. The only source of gold shown in the Certificates is the <0.5 ppm listed for the copper rods used to make the electrodes. Again, the question whether the experimental process is capable of increasing the concentration of Au from <0.5 ppm as reported in the copper prior to experiment to the 162 ppm detected after the experiment awaits investigation. The third possibility, one that is not accepted by modern science, is that at least three low energy nuclear reactions took place during the experiment.

Placing the sulfur-lithium test material in contact with the lead anode, pumping down to vacuum, admitting oxygen as a catalyst, and electrifying the tube may have caused the electrical repulsive force existing between two plus charged nuclei—lithium and sulfur—to become temporarily neutralized. Such temporary neutralization may have allowed the omnipresent centripetal Casimir force to cause a nucleus of lithium-7 and a nucleus of sulfur-32 to react and form potassium-39. As the fusion process took place, a simultaneous fission reaction occurred, in which the highly saturated and unstable lead anode fractured and surrendered atoms of lithium ($^{204}Pb - {}^{7}Li \rightarrow {}^{197}Au$), or lead-204 minus lithium-7 $\rightarrow$ gold-197. The equation can also be written as:

$$^{204}Pb \rightarrow {}^{7}Li + {}^{197}Au$$

Meanwhile, in another region of the periodic table, lithium may have reacted with copper to form germanium ($^{63}Cu + {}^{7}Li \rightarrow {}^{70}Ge$) All in all an intriguing set of possibilities.

Source: Edward Esko, Anomalous Metals in Electrified Vacuum, *Infinite Energy* Issue 99, Sept./Oct. 2011.

9

IN SEARCH OF THE PLATINUM GROUP

ABSTRACT

In studies sponsored by the New Energy Foundation (NEF) and conducted at Quantum Rabbit (QR) lab in Owls Head, Maine, on December 17, 2009, a QR research team including Alex Jack, Woody Johnson, and me, assisted by Bill Zebuhr and vacuum consultant Steve Hansen, conducted vacuum discharge tests utilizing pure graphite rods in combination with strontium and sulfur test materials. The goal of these studies was to see if we could trigger low energy nuclear reactions in which strontium would react with carbon and with atmospheric fill gases to form ruthenium and other platinum group metals.

BACKGROUND

Ruthenium is one of the rarest elements on earth. It is the first of the precious metals, and is part of the platinum group of metals, occurring naturally in ores in which platinum is found. Earlier studies at QR labs suggested the possibility of inducing low energy nuclear reactions under relatively low temperatures and pressures, and with relatively small inputs of energy. In previous studies, ICP (Inductively Coupled Plasma Atomic Emission Spectroscopy) analysis of Quantum Rabbit test samples had noted the anomalous appearance of copper, palladium, strontium, chromium, germanium, and tin on electrodes and test materials used in QR vacuum studies.

These results point to the possibility of low energy nuclear reactions, in which lighter elements react, under relatively low temperature, pressure, and energy, and create heavier elements.

METHOD

Like previous QR vacuum studies, the Dec. 17 tests took place in a specially designed vacuum tube. The borosilicate tube was 150 mm long, with a 60 mm quartz midsection. A 3/8-inch diameter quartz straight section, perpendicular to the tube and 75 mm in length, connected the quartz reaction zone with the vacuum manifold, and served as the entrance for backfill from the surrounding atmosphere. We had ordered high purity graphite rods from Alfa Aesar and used the rods to create the anode and cathode. The graphite electrodes were inserted into aluminum electrodes, 0.38-inch in diameter, designed to be flush with the inner wall of the tube so as to seal the tube to maintain vacuum. The aluminum electrodes were positioned outside the quartz reaction zone so as not to interfere with reactions taking place in the center of the tube. The graphite rods were ultra high purity, 99.9995% pure (metals basis). Their C of A is as follows (Stock 14754/Lot F27S027):

Graphite electrode, counter-flat top, 3.05mm dia, 38.10 mm long, 99.9995% (metals basis)*

Al	ND**	Fe	ND	Ni	ND	W	ND
B	ND	Pb	ND	Si	ND	V	ND
Ca	ND	Mg	ND	Ag	ND	Zn	ND
Cr	ND	Mn	ND	Sn	ND	Zr	ND
Cu	ND	Mo	ND	Ti	ND		

*Values given in ppm unless otherwise noted
**ND: Not detected

First, the graphite cathode was secured in the tube, and a catalyst of pure sulfur placed in the quartz midsection. The sulfur, listed in the Alfa

Aesar catalogue as "sulfur pieces," was actually a coarse powder. It was labeled as Puratronic® (Stock 10755/Lot D03S015) and certified as being 99.999% pure (metals basis.) Like the graphite rods, the sulfur powder contained no trace of ruthenium or any other platinum group metal. After the sulfur was placed in the tube, a piece of strontium was positioned in the center of the reaction zone. The strontium, listed in the Alfa Aesar catalogue as "strontium granules," had started to oxidize and had a thin grey coating on the surface. The C of A from Alfa Aesar for the strontium granules showed no trace of ruthenium or any other platinum group metal (Stock 35789/Lot G25R040.)

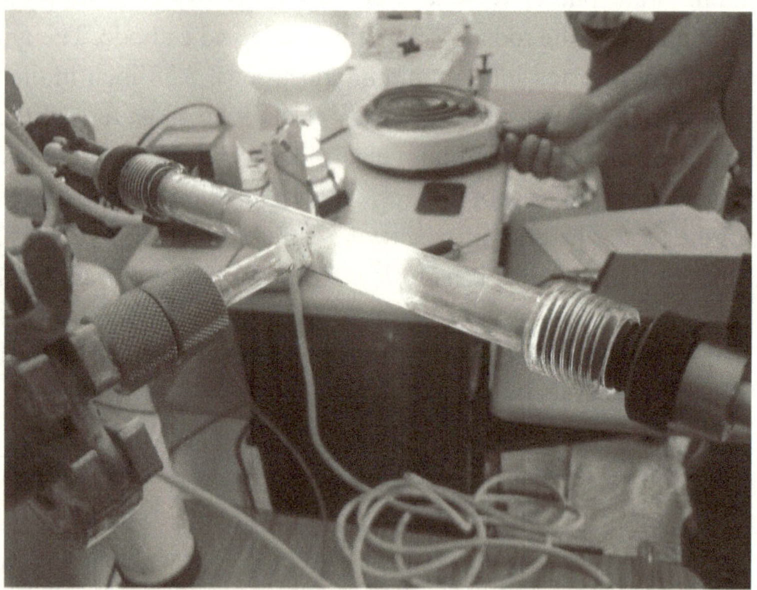

Strontium-sulfur plasma in the QR tube, Dec. 17, 2009

Following insertion of the sulfur and strontium, the graphite an ode was inserted in the opposite end of the tube, flush with the strontium piece. The electrodes were connected to the power supply. The tube was pumped down to vacuum and backfilled with pure oxygen to approximately 3.5 torr. Direct current (DC)was passed through the electrodes, producing an electric arc and glow discharge with the blue/white characteristic of sulfur. After about two minutes, the current was lowered to a level sufficient to maintain the arc. This was sustained for about five minutes, at which time the current was boosted to the level sufficient to spark an active discharge. This alternating high-low-high-low pulse was maintained for over twenty minutes. Also, during this period, electrode polarity was reversed, with anode becoming cathode and cathode becoming anode. After the test period, the electricity was turned off, the vacuum pumps disconnected, and the sample allowed to cool.

The test was then repeated, with another set of graphite electrodes and strontium and sulfur test materials. The protocol was essentially the same with several slight differences. In Test 2, the strontium test material was placed in contact with the graphite cathode rather than the anode. Also, during the approximately twenty minute test period, we varied the amount of pressure in the tube, with a low of 3.5 torr and a high of 30 torr. During the period of slightly elevated pressure, we noticed that the strontium vapor glowed brightly with a reddish-violet hue.

RESULTS

Three samples were retrieved from each test: the graphite anode, cathode, and strontium-sulfur residue. The samples were packaged and sent to New Hampshire Materials Laboratory for ICP analysis. ICP results came back on January 7, 2010. In addition to ruthenium, we asked NHML to scan for other platinum group metals including rhodium, palladium, and platinum. ICP analysis of the test materials came back with the same values for all six samples (NHML File 27109/January 7, 2010):

Element	**Test 1 and 2 Graphite Electrodes/Sr Residue***
Ruthenium	<26 ppm
Palladium	<100 ppm
Rhodium	<32 ppm
Platinum	<5 ppm

*Values in parts per million (ppm)

Due to the small size of the samples, ranging from a low of 0.0733 gm to a high of 1.5419 gm, none of the requested elements was found down to the detection limits listed in the report. Below the detection limits it was impossible to determine whether or not the requested element was present. In other words, the results were inconclusive. It was impossible to state for certain whether we did or did not achieve the sought-after low energy transmutations.

ANALYSIS

As mentioned above, we had selected ruthenium as the target metal for the Dec. 17 tests. Our hope was to prove the formula:

$$^{88}Sr + {}^{12}C \rightarrow {}^{100}Ru$$
Strontium-88 + carbon-12 → ruthenium-100

In addition, we had speculated that a variety of other low energy transmutations would take place involving the elements used in the experiment, including nitrogen and oxygen from the open air. These low energy reactions included:

$$^{88}Sr + {}^{15}N \rightarrow {}^{103}Rh$$
Strontium-88 + nitrogen-15 → rhodium-103

$$^{88}Sr + {}^{16}O \rightarrow {}^{104}Pd$$
Strontium-88 + oxygen-16 → palladium-104

Several factors may have influenced the inconclusive result, including:

1. Failure to reach high enough temperature. All of the platinum group metals have very high melting points. It may be necessary to achieve temperatures close to or above these points to facilitate low energy transmutation. The glass vacuum tube used in the experiment may not be able to sustain such high temperatures. A new tube design may need to be employed for this purpose.

2. Oxidation of the strontium test sample. The oxide on the surface of the strontium pieces may have interfered with the availability of pure strontium for a low energy nuclear reaction. Future tests should use fresh strontium with no oxide coating. Due to the oxidation of the strontium test sample, the QR team was essentially dealing with strontium oxide rather than pure strontium. A brief comparison between the two may shed light on the result achieved in the QR experiment.

	Strontium	**Strontium Oxide**
Melting Point	777 degrees C	2531 degrees C
Boiling Point	1382 degrees C	>3000 degrees C

A brief comparison between the two may shed light on the result achieved in the QR experiment. The higher vaporization point of strontium oxide versus pure strontium no doubt played a role in the outcome of the test.

3. Use of atmosphere as fill gas rather than pure oxygen. Previous tests in which a definitive result was achieved employed pure oxygen as the fill gas. The use of atmosphere (nitrogen/oxygen) as fill rather than pure oxygen may have impeded potential low energy nuclear reactions.

It is important to note that we have received similar results in past tests, only to see the predicted formulas apparently proven in subsequent studies. In tests conducted on Sept. 30, 2008, I had predicted the appearance of germanium-70 on copper-63 electrodes following a low energy reaction with lithium-7. The analysis of the Sept. 30 test materials came back inconclusive, with the detection limits reached on all samples. However, in subsequent tests, germanium was found on test samples at up to 416 ppm (on Dec. 30, 2008) and 3196 ppm (on July 30, 2009). See "Appearance of Tin on a Silver Anode," *Infinite Energy* Issue 88. Whether any of the platinum group metals will appear with certainty in repetitions of the above test remains to be seen.

Further research is of course needed to prove or disprove these assertions. Rather than being definitive, Quantum Rabbit research has been *suggestive* of the possibility of low energy nuclear reactions. It is hoped that additional research will expand our knowledge of this phenomenon and open the possibility of a new paradigm of scientific understanding and sustainable technology.

Source: Edward Esko, "In Search of the Platinum Group Metals," *Infinite Energy* Issue 92, July/August 2010.

10

ANOMALOUS METALS, PART II

ABSTRACT

In a study funded by the New Energy Foundation and conducted at Quantum Rabbit (QR) lab in Owls Head, Maine, on September 27, 2011, independent analysis by inductively coupled plasma spectroscopy (ICP) of test samples revealed the anomalous appearance of potassium (K) at 181 ppm (parts per million) and gold (Au) at 252 ppm. The vacuum discharge test employed a copper cathode, pure lithium and sulfur test material, and a pure copper anode at the center of which a pure lead insert had been pressed. Scientific grade neon was added to the vacuum tube to strike plasma, followed by a catalyst of pure oxygen. Although it is possible that test materials were contaminated, the appearance of these anomalous metals once again raises the possibility of low energy transmutation.

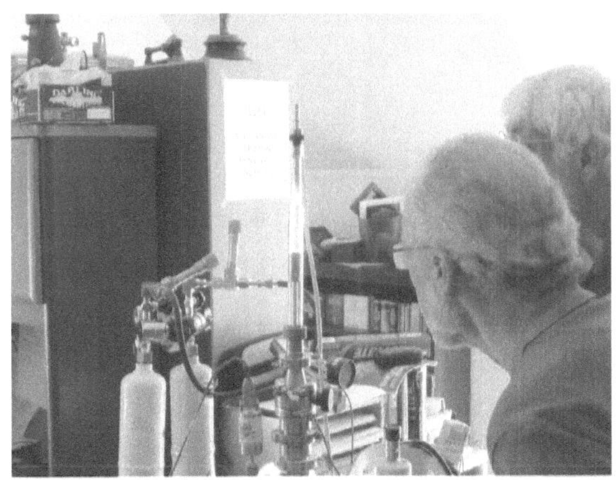

Bill Zebuhr and Steve Hansen monitor the 2011 experiment

BACKGROUND

The 2011 test followed a series of tests conducted at the QR lab in Owls Head on July 30, 2009, and described in my paper "Anomalous Metals in Electrified Vacuum" published in *Infinite Energy*, Issue 99. In the 2009 tests, a lead insert was pressed into a copper anode. A copper cathode was inserted into one end of the vacuum tube, followed by pure lithium and sulfur test material. The lead-tipped anode was then inserted, the tube pumped down to vacuum, and pure oxygen admitted to approximately 3.5 torr. An electric arc was struck and a glow discharge with the characteristic color of lithium was produced. ICP analysis of test samples revealed the anomalous presence of germanium (Ge) at up to 3196 ppm; potassium (K) at up to 750 ppm; and gold (Au) at up to 174 ppm. Once again, aside from possible minute traces listed in the Certificates of Analysis of several test components, none of the anomalous metals was introduced into the experiment.

VACUUM DESIGN

The 2011 test employed a vacuum tube with a different design than that used in 2009. In the 2009 tests, a vacuum tube made of borosilicate was positioned horizontally atop the vacuum manifold and attached to the manifold by a pump out port. The tube contained a quartz midsection where the reactions took place. The 2011 tube was made entirely of quartz and fastened vertically atop the vacuum manifold. This eliminated the perpendicular joint that connected the earlier tube with the manifold. It was our calculation that this new design would tolerate higher temperatures with less risk of breakage.

THE EXPERIMENT

The 2011 test employed the same inputs as the 2009 experiments, with several new features. The first modification was the creation of a small recess at the center of the anode. The recess facilitated a more secure placement of test material in the tube. The recess helped confine the test material to the reaction zone. As was the case in 2009, a lead insert was placed in the center of the copper anode. The lead insert consisted of a lead slug approximately 0.25-inch diameter pressed into a 0.25-inch by 0.25-inch drilled hole. One piece of lithium was centrally placed atop the lead insert. The lithium was surrounded by sulfur pieces. Electrode separation was adjusted to a minimum value with just enough clearance to reduce shorting. Typically this was in the range of 0.30 to 0.60 inches. The second modification was the use of neon as a fill gas to strike plasma before admitting oxygen as the catalyst. (Oxygen was the sole fill gas in the 2009 tests.)

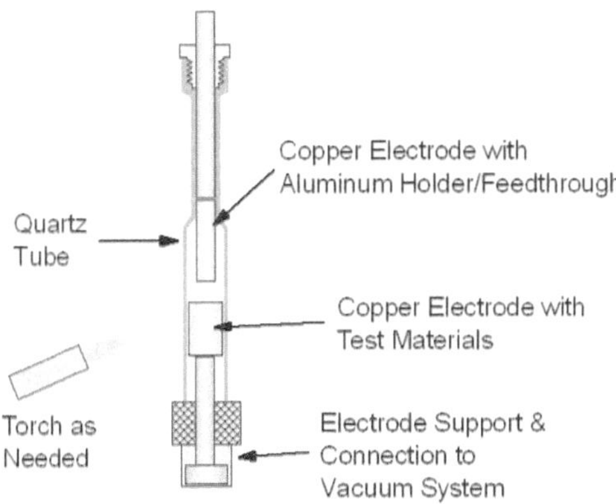

General QR tube configuration showing copper cathode assembly, copper anode, quartz tube, and vacuum system connection. The torch was not used in the test

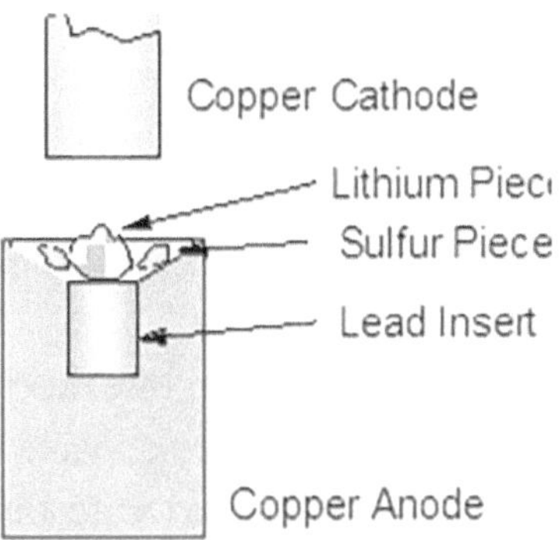

Electrode configuration and test material

In the worksheet that I drafted prior to the test, I summarized the protocol as follows:

Inputs:
Copper electrodes
Lead insert in anode
Lithium test material
Sulfur test material
Neon fill
Oxygen fill

Procedure:

1. Insert is placed into the anode.
2. Measured quantity of test materials are placed in recess.
3. Glass/quartz tube is placed over the anode assembly.
4. Cathode is inserted into the tube and secured at the desired separation from the anode.
5. Fill with neon to 2 torr.
6. Strike plasma using direct current (DC).
7. Heat reaction zone with hand torch (optional: this was not done in the experiment).
8. Maintain conditions for 2 minutes.
9. Admit oxygen fill to 6 torr. Continue until reaction noticeably slows or tube is in danger of breaking.
10. Disconnect power and allow sample to cool.

The test proceeded in real time as follows (keep in mind that the data points are approximate.) The tube was pumped down to vacuum. Neon was admitted at the start to 2 torr. At 1 minute in, the torr reading was 3.0, while power supply readings measured 53 volts and 6.95 amps. The inside of the tube was glowing red-purple, with what appeared to be the color of neon plasma. At about 2 minutes, the readings were as follows: 3.0 torr, 70 volts, and 5.63 amps. Intense heat was generated at this point; so that the test materials appeared to be melting. Oxygen was admitted between minute 3 and 4, and the torr reading went up to approximately 8.28. Following the oxygen fill, the test material began glowing a ruby red, the characteristic color of lithium plasma.

At around 6 minutes, there was concern that the tube had failed. Power was disconnected. Thirty seconds later it was decided that the tube was still viable, and the decision was made to admit fresh oxygen and fire up the tube once again. At this point the tube began glowing blue-green. Between 8 to 9 minutes, the power alternated between 45-55 volts and 6.65 and 7.5 amps. After 10 minutes, conditions in the tube appeared to stabilize. Voltage hovered around 77 and amps around 5.4. The test finished after a total time of approximately 14 minutes with a 3.5 torr reading. Upon conclusion of the experiment, the electricity was turned off, the vacuum pumps disconnected, and the samples allowed to cool.

The QR Power Supply

The DC (Direct Current) power supply is based on a microwave oven transformer. These transformers are current limited. With a 110 volt mains input the output will be about 3500 volts. As the transformer is loaded, the output voltage will become reduced to maintain a constant output current.

The "hot" output of the transformer is connected to a diode to provide a half-wave rectified output. Since the full output power of the transformer (about 1000 watts) could quickly overheat the QR experiment tube, the primary is ballasted with a high current series resistance consisting of a 600-watt heater element in parallel with a 150-watt incandescent lamp.

For monitoring the power supply we did not measure the actual tube voltage or current through the tube. Instead we monitored the primary voltage and the current into the primary. Voltage was measured with a digital volt meter (DVM) and the current with a commonly available "Kill-A-Watt" meter.

With no load on the transformer, the primary voltage would be close to the mains voltage (118 volts rms is normal here) with the current in the 3-4-amp range. With the tube in an arc condition (high current, low discharge voltage) the primary voltage would drop and the current would increase. The typical arc voltage would be in the 1-volt range vs. glow discharge in the range of a few hundred volts. Note that when the tube is conducting, the voltage across the tube will always be significantly lower than the approx. 3500-volt open circuit voltage. Multiplying the voltage times the current, about 300 V-A in most cases, may approximate power being delivered to the tube. **—Steve Hansen, QR Vacuum Consultant**

RESULTS AND ANALYSIS

Two sets of samples were retrieved for testing: the lead-tipped copper anode with lithium-sulfur residue in its center, the copper cathode, and the quartz tube itself, which contained residue on its inner surface. The anode, cathode, and inside of the tube had undergone noticeable changes during the experiment. The samples were carefully packaged and sent to New Hampshire Materials Laboratory for ICP analysis. The Test Report came back on October 14, 2011 (NHML File Number 28929) and is reproduced below.

New Hampshire
MATERIALS
Laboratory, Inc.

TEST REPORT

October 14, 2011
File Number: 28929
Mr. Edward Esko
Quantum Rabbit LLC

Overview:
Samples Received: QR test samples
Work Requested: Elemental analysis
Sample Disposition: Consumed by analysis

Analysis Results:

Test 1	**Anode/Residue**	**Cathode/Tube**
Potassium	181 ppm	<0.01 ppm
Gold	252 ppm	<0.01 ppm
Sample weight (gm)	0.2083	1.7429

Prepared by:
Timothy M. Kenney
Director of Laboratory Services

The 2011 results paralleled the results of 2009. Two of the three anomalous metals that appeared in 2009 were detected in 2011: potassium (K) at 181 ppm and gold (Au) at 252 ppm. Germanium, the third anomalous metal to appear in 2009, was not detected in the 2011 test. A comparison between the 2009 test and the 2011 experiment is presented below, followed by an analysis of each of the 2011 test components taken from their Certificates of Analysis.

Comparison of 2011 and 2009 Experiments		
Element	**Starting Concentration (ppm)***	**Final Concentration (ppm)****
2011		
Au	<0.51	252
K	<1.615	181
*2009****		
Au	<0.5	174
K	36.610	750

*Certificate of Analysis provided by Alfa Aesar; additional 2009 value for K provided by M & M Glassblowing.
**ICP Analysis by New Hampshire Materials Laboratory.
***Test 1 conducted on July 30, 2009. Au and K appeared in two subsequent tests on July 30.

Analysis of 2011 Inputs*		
Input	**K (ppm)**	**Au (ppm)**
Copper rod (anode)		
Stock: 12754/Lot: J22P23	<0.01	<0.01
Copper rod (cathode)		
Stock: 10156/Lot: E18U005	<0.005	<0.5
Lead pellet		
Stock: 43415/Lot: A10X038	ND**	ND

Lithium piece		
Stock: 10773/Lot: H17W051	<1	ND
Sulfur pieces		
Stock: 10755/Lot: A06X003	ND	ND
Quartz tube***	0.6	ND

*Certificate of Analysis provided by Alfa Aesar.
**None detected
***Analysis provided by M & M Glassblowing.

The 2011 results raise the question that was raised in 2009, namely, where did the anomalous metals come from? In the first "Anomalous Metals" article, I presented three possible sources for the anomalies: 1) contamination, or the presence of anomalies in test materials prior to experiment; 2) concentration, or the gathering of anomalous elements in the tested region; and 3) transmutation, or the formation of anomalies through low energy nuclear reactions. The third possibility, that the anomalies appeared through a process of low energy transmutation, is not accepted by modern science. To glean more background, we asked NHML to do ICP analysis on a control sample of copper taken from the actual batch used in the test, together with a sample of lead from the test batch. The samples were sent to NHML at the end of October and the Test Report, File 29008 presented below, came back on November 3.

New Hampshire
MATERIALS
Laboratory, Inc.

TEST REPORT

November 3, 2011
File Number: 29008

Overview:
Samples Received: (1) Lead Pellet, (2) Copper Rod

Work Requested: Elemental analysis for Au
Sample Disposition: Consumed by analysis

Analysis Results:

<u>Sample</u>   <u>Au Content</u>

Lead Pellet <1 ppm

Copper Rod <1 ppm

Prepared by:
Timothy M. Kenney
Director of Laboratory Services

I also requested that each of our vendors reply to the question of whether their equipment or handling of our samples could have led to "contamination" with Au. Steve Hansen, our vacuum system designer and consultant replied: "I am not aware of any way that Au could be introduced. Au is not part of any portion of the system and nothing in the system has been exposed to Au, save the Au that has appeared in the tests. None of the machine tools have Au as a component, etc." Wayne Martin, of M & M Glassblowing, who fabricated the vacuum tube used in the test answered: "Regarding the Au, I don't see any trace amounts listed in the chemical composition of any of the glasses used to make the tubes. So I guess I can't explain it." Tim Kenney, who conducted the ICP analysis of our samples at New Hampshire Materials Lab stated: "We rarely deal with Au, and had nothing else in house when your samples were analyzed. Accordingly, the detected Au should not have come from us."

SUMMARY

In the first "Anomalous Metals" article I offered a possible pathway for the transmutation process: "Placing the lithium-sulfur test material in contact with the lead anode, pumping down to vacuum, admitting oxygen as a catalyst and electrifying the tube, may have caused the electrical repulsive force existing between two positively charged nuclei—lithium and sulfur—to become neutralized.... thus allowing the centripetal Casimir force to cause lithium-7 nuclei to fuse with sulfur-32 to produce potassium-39.

As fusion took place, a fission reaction occurred in which the lead anode fractured and surrendered nuclei of lithium (^{204}Pb − ^{7}Li → ^{197}Au.)"

In other words, according to the quantum conversion hypothesis, the process of cool fusion, in which lithium and sulfur fuse to form potassium, initiates a process of cool fission, in which nuclei of lithium are subtracted from nuclei of lead: Li + S → K (*fusion reaction*) followed by Pb → Li + Au (*fission reaction*). Such quantum events, if they are eventually proven to occur, seem to be taking place in the realm of subatomic particles, thus the purported transmutation products are recorded in parts per million, not in ounces or grams.

Because these events have so far been limited to a minute scale, the specter of contamination, always lurking in the background, is difficult to rule out, our best efforts notwithstanding. Whether or not such quantum events can be replicated on a scale large enough to discount contamination remains to be seen. It also remains to be seen whether such events can be made to occur with regularity and precision, or like the events observed here, occur with a certain degree of randomness and unpredictably.

In the realm of theory, one scientific reviewer—a trained physicist— has suggested I not attempt to explain these (and other) possible reactions in terms of the Casimir force. As he stated in an email: "It would be better to say that currently there is no explanation for the suggested transmutation reactions, but if transmutation could be confirmed it will be necessary to revise the existing laws of physics." However, the phenomenon known as "quantum tunneling" may offer an alternative to the Casimir explanation. In my letter "Quantum Tunneling and the Quantum Rabbit Effect," published in *Infinite Energy* (July 2010) and reprinted in *Cool Fusion*, I propose that under certain conditions, such as those present in QR vacuum tubes, nuclei act more like waves and less like particles. In this model, the nucleus be- haves not as a classical particle but as a quantum wave that obeys the laws of quantum tunneling followed by electron waves. If quantum tunneling applies to nuclei in the way it applies to electron wave-packets, it could explain how a small fraction of positively charged nuclei are able to tunnel through, go around, or somehow breach the Coulomb barrier and achieve fusion with other positively charged nuclei, thus producing the anomalies noted by QR.

The reviewer raises another question: "You have to put a huge amount of energy into the reaction in order to split Pb-204 into Li-7 and Au-197. Where would that energy come from?" Perhaps the energy for the fission reaction comes from the fusion reaction in which lithium combines with sulfur to form potassium. In other words, in this experiment, the fusion of Li-7 and S-32 may have produced a quantum of energy sufficient to trigger the fission of lead into the two lighter elements suggested above.

Source: Edward Esko, "Anomalous Metals Part II," *Infinite Energy* Issue 103, May/June 2012.

11

IN SEARCH OF THE PLATINUM GROUP, PART II

ABSTRACT

In a study funded by the New Energy Foundation and conducted at Quantum Rabbit (QR) lab in Owls Head, Maine, on September 27, 2011, independent analysis of test samples by inductively coupled plasma spectroscopy (ICP) revealed the anomalous appearance of aluminum (Al)), scandium (Sc), and selenium (Se). The vacuum discharge test employed a copper cathode, pure boron and sulfur test material, and a pure copper anode at the center of which an insert of pure bismuth had been pressed. Scientific grade neon was added to the vacuum tube to strike plasma, followed by a catalyst of pure oxygen. Although it is possible that test materials were contaminated, the appearance of these anomalies raises the possibility of low energy transmutation.

BACKGROUND

The 2011 test followed a series of tests conducted at the QR lab in Owls Head on July 30, 2009, and described in my paper "Anomalous Metals in Electrified Vacuum" published in *Infinite Energy*, Issue 99. In the 2009 tests, a lead insert was pressed into a copper anode. A copper cathode was inserted into one end of the vacuum tube, followed by pure lithium and sulfur test material. The lead-tipped anode was then inserted, the tube pumped down to vacuum, and pure oxygen admitted to approximately 3.5 torr. An electric arc was struck and a glow discharge with the characteristic color of lithium was produced. ICP analysis of test samples revealed the anomalous presence of germanium (Ge) at up to 3196 ppm (parts per million); potassium (K) at up to 750 ppm; and gold (Au) at up to 174 ppm. Once again, aside from possible minute traces (mostly below the limits of detection) listed in the Certificates of Analysis of test components, none of the anomalous elements was introduced into the experiment.

In the article I discuss the possibility that the anomalies were produced through a process of low energy transmutation, specifically, a cool fission reaction (Pb – Li → Au) triggered by a cool fusion reaction (Li + S → K). The 2011 experiment was designed to test the cool fusion → cool fission hypothesis, this time using bismuth rather than lead, and boron in place of lithium, the idea being to see if it would be possible to fission bismuth-209 into platinum-198:

$$^{209}Bi \rightarrow {}^{11}B + {}^{198}Pt$$

This was QR's third attempt at achieving platinum group metals. The first, conducted in September 2008, saw the anomalous appearance of palladium (Pd) on a zinc (Zn) anode, a possible confirmation of the formula: $^{68}Zn + {}^{34}S \rightarrow {}^{102}Pd$. The experiment ("Appearance of Palladium on a Zinc Anode) was published in *Infinite Energy* Issue 87, September/October, 2009. The second effort, a series of experiments designed to produce ruthenium (Ru), was tried in December 2009 and was inconclusive (see "In Search of the Platinum Group Metals, " *Infinite Energy*, Issue 92, July/August 2010.)

Vacuum Tube Design

In the 2008 and 2009 tests, a vacuum tube made of borosilicate was positioned horizontally atop the vacuum manifold and attached to the manifold by a pump out port. The tube contained a quartz midsection where the reactions took place. The 2011 tube was made entirely of quartz and fastened vertically atop the vacuum manifold. This eliminated the perpendicular joint that connected the earlier tube with the manifold. It was our calculation that this new design would tolerate higher temperatures with less risk of breakage.

The Experiment

In addition to the vertical tube, the 2011 test differed from the earlier experiments in several respects. The first was the design of a small recess at the center of the anode. The recess facilitated a more secure placement of test material in the tube. The recess helped confine the test material to the reaction zone. To create the insert, small round beads of pure bismuth, 1-5 mm in size, were poured into a 0.25-inch by 0.25-inch diameter hole and pressed to bind them as one. Boron powder, in the form of tiny crystals -4+40 mesh, was sprinkled atop the bismuth at the center of the anode. Sulfur pieces were then added. Electrode separation was adjusted to a minimum value with just enough clearance to reduce shorting. Usually this was in the range of 0.30 to 0.60 inches. The second modification was the use of neon as a fill gas to strike plasma before admitting oxygen as the catalyst. (Oxygen was the sole fill gas in the earlier tests.) In the worksheet drafted prior to the test, I summarized the procedure as follows:

Inputs:
Copper electrodes
Bismuth insert in anode
Boron test material
Sulfur test material
Neon fill
Oxygen fill

Procedure:
1. Insert is placed into the anode.
2. Measured quantity of test materials are placed in recess.
3. Glass/quartz tube is placed over the anode assembly.
4. Cathode is inserted into the tube and secured at the desired separation from the anode.
5. Fill with neon to 2 torr.
6. Strike plasma using direct current (DC).
7. Heat reaction zone with hand torch (optional: this was not done in the experiment).
8. Maintain conditions for 2 minutes.
9. Admit oxygen fill to 6 torr. Continue until reaction noticeably slows or tube is in danger of breaking.
10. Disconnect power and allow sample to cool.

In the lab, the test lasted for approximately 10 minutes. Power remained fairly constant at around 50 volts and 7 amps. From the initial 2 torr pressure, the alternating neon and oxygen fill averaged 3.5 torr. At one point, the pressure was increased to 11.9 torr to generate a stronger arc, and then returned to 3.5 torr. Upon completion of the experiment, the power was disconnected and the test sample allowed to cool.

RESULTS

Two sets of samples were retrieved for testing: 1) the copper anode with the bismuth insert and boron-sulfur residue at the center, the copper cathode; and 2) and the quartz tube itself, which contained residue on its inner surface. The anode, cathode, and inside of the tube had undergone noticeable changes during the experiment. The samples were carefully packaged and sent to New Hampshire Materials Laboratory for ICP analysis. The Test Report came back on October 14, 2011 (NHML File Number 28929.)

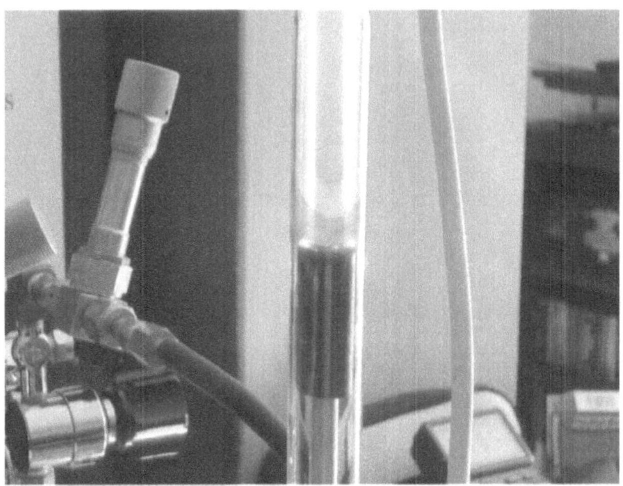

The QR tube with boron-sulfur test material (center)
atop the bismuth-filled anode (below)

ICP analysis revealed the anomalous appearance of three elements: scandium (Sc) at 18 ppm, aluminum (Al) at 342 ppm, and selenium at 12 ppm. Platinum was either not present, or was present in quantities too small to be detected. A comparison of the quantities of scandium, aluminum, and selenium reported in the Certificates of Analysis of the test materials (Starting Concentration), followed by the quantities of these elements discovered by ICP following the experiment (Final Concentration), appears after the Test Report. Also presented is a table presenting a breakdown of the quantities of the anomalous elements found in each experimental input as reported in the Certificates of Analysis provided by Alfa Aesar, supplier of the materials used in the test. The analysis of the quartz tube material, provided by M & M Glassblowing, the tube fabricator, is also shown.

New Hampshire
MATERIALS
Laboratory, Inc.

TEST REPORT

October 14, 2011
File Number: 28929
Mr. Edward Esko
Quantum Rabbit LLC

Overview:
Samples Received: QR test samples
Work Requested: Elemental analysis
Sample Disposition: Consumed by analysis

Analysis Results:

Test 3	Anode/Residue	Cathode/Tube
Platinum	<0.01 ppm	<0.01 ppm
Scandium	3 ppm	15 ppm
Aluminum	62 ppm	280 ppm
Selenium	<0.01 ppm	12 ppm
Sample weight (gm)	0.3342	0.3334

Prepared by:
Timothy M. Kenny
Director of Laboratory Services

ANALYSIS

In previous papers I discussed three possible explanations for anomalies such as these: 1) contamination, or the presence of anomalies in test materials prior to experiment; 2) concentration, or the congregation of anomalous elements in the tested region; and 3) transmutation, or the formation of anomalies through low energy nuclear reactions. In this paper, we shall limit ourselves to the third possibility.

Anomalies in the Sept. 27 Test		
Element	Starting Concentration (ppm)*	Final Concentration (ppm)**
Sc	<0.1012	18
Al	16.143	342
Se	<0.3087	12

*Certificate of Analysis provided by Alfa Aesar; quantities below detection limits included in totals.

**ICP Analysis by New Hampshire Materials Laboratory.

Analysis of Inputs*			
	Sc ppm	Al ppm	Se ppm
Copper rod (anode) Stock: 12754/Lot: J22P23	<0.001	0.004	<0.01
Copper rod (cathode) Stock: 10156/Lot: E18U005	<0.0002	<0.021	<0.005
Bismuth beads Stock: 38619/Lot: B26W020	<0.1	<0.1	<0.1
Boron powder Stock: 10112/Lot: F05U009	ND**	0.018	<0.0037

Sulfur pieces			
Stock: 10755/Lot: A06X003	ND	ND	<0.1
Quartz tube***	ND	16	ND

*Certificate of Analysis provided by Alfa Aesar.
**None detected
***Analysis provided by M & M Glassblowing.

In our quantum conversion hypothesis, the process of cool fusion can be used to trigger a process of cool fission. In 2009, we tested this hypothesis with lead (atomic number 82) with intriguing results. In the 2011 tests, we were attempting to test the hypothesis with bismuth (atomic number 83) the element that follows lead on the periodic table, and the heaviest of the non-radioactive elements. In 2011 we attempted to subtract boron-11 from bismuth-209 (*fission reaction*) to produce platinum-198 (^{209}Bi $\rightarrow$ ^{11}B + ^{198}Pt). In order to accomplish this, a reaction in which boron-11 fused with oxygen-32 and/or sulfur-34 would be required. In theory, the fusion reactions would result in the creation of aluminum-27 (^{11}B + ^{16}O $\rightarrow$ ^{27}Al) and/or scandium-45 (^{11}B + ^{34}S $\rightarrow$ ^{45}Sc).

If all three reactions took place as predicted, in addition to aluminum and scandium, the test sample would have contained platinum. As we can see from the Test Report, platinum was not found in detectable quantities, even though the other predicted metals were noticed. We can interpret this to mean that the fusion reactions were either not strong enough to fission bismuth, or if they were strong enough, the resulting platinum was too insignificant to detect. Meanwhile, as predicted, a secondary reaction seems to have taken place. A small portion of the copper-65 in the electrodes may have fused with boron-11 to form selenium-76: ^{65}Cu + ^{11}B $\rightarrow$ ^{76}Se. [1] The prevalence of the various isotopes in the raw test materials may explain why certain anomalies seem to be more prevalent than others in the finished samples.

In the reaction ^{11}B + ^{16}O $\rightarrow$ ^{27}Al, we see that boron-11 makes up 80.1% of naturally occurring boron isotopes, and oxygen-16 makes up 99.8% of natural oxygen isotopes. In contrast, in the reaction ^{11}B + ^{34}S $\rightarrow$ ^{45}Sc, sulfur-34 comprises only 4.2% of the naturally occurring isotopes. At 99.8%, oxygen-16 is slightly more than 23 times more widespread amongst oxygen atoms than sulfur-34 is amongst sulfur atoms, thus, the fusion of oxygen-16 with boron-11 yielded 342 ppm aluminum-27, a little more than 20 times the 18 ppm scandium-45 produced by the hypothetical fusion of boron-11 and sulfur-34. There was far more oxygen-16 available for the reaction than there was sulfur-34, a possible explanation for the higher incidence of aluminum-27 in the result.

Footnotes:

[1] The test was repeated with similar results (Test 4 NHML File Number 2892). Scandium was reported at 9 ppm, aluminum at 164 ppm, and selenium at 5 ppm. In addition, lithium was introduced in that experiment. Accordingly, sodium was reported by ICP analysis at 50 ppm; a result of the possible fusion of lithium-7 with oxygen-16 ($^7Li + {}^{16}O \rightarrow {}^{23}Na$.)

Source: Edward Esko, "In Search of the Platinum Group Part II," *Infinite Energy* Issue 104, July/August 2012.

12

Quantum Tunneling and the Quantum Rabbit Effect

I would like to respond to the article, "Was Transmutation Observed at the Quantum Rabbit Laboratory," *Infinite Energy* Issue 92, July/August 2010, in which Matthias Grabiak states that the Coulomb barrier between positively charged protons is too strong to allow low energy fusion to take place. As he puts it: "Conventional physics tells us that it requires pressures and temperatures comparable to the interior of the sun to overcome the Coulomb barrier even for the very lightest elements to achieve fusion. But, for fusing heavier elements as suggested by the Quantum Rabbit experiments, only cataclysmic scenarios like supernova explosions will suffice."

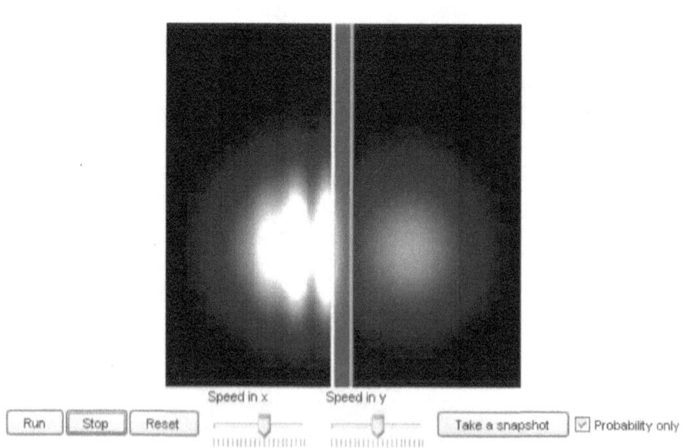

It may be that the phenomenon of quantum tunneling offers an explanation of how the Coulomb barrier can be breached with relatively low inputs of temperature, pressure, and energy and how nuclear transmutation can be achieved under these conditions. In the above illustration we see the reflection and tunneling of an electron wave packet directed at a potential barrier. The bright spot moving to the left is the reflected part of the wave packet. The dim spot moving to the right is the small fraction of the wave packet that tunnels through the classically forbidden barrier.

Applying this phenomenon to the QR experiment in which the possible low energy transmutation of lithium into potassium was noted ("Appearance of Potassium in a Li-S Matrix," *Infinite Energy* Issue 91), the bright spot on the left corresponds to the lithium nuclei used in the test. The potential barrier represents the Coulomb barrier existing between two positively charged nuclei, in this case, those of lithium and sulfur used in the experiment. In this model, a small fraction of the lithium nuclei, which exist in the form of a wave packet (represented by the dim circle on the right), tunnels through the barrier and moves near enough to the sulfur nuclei for the strong force to bind them together to form larger nuclei of potassium (7Li + 32S → 39K). Please note that in my articles I refer to the strong force as the centripetal Casimir force found throughout the universe.

In this model, the nucleus behaves not as a classical particle but as a quantum wave that obeys the same laws of quantum tunneling as an electron wave.

The diagram below offers a schematic version of this possibility. In the experiment described above, the wave to the left of the barrier (vertical line) represents the total number of lithium nuclei approaching the Coulomb barrier in the form of a wave packet. The wave to the right of the barrier in the lower diagram on the next page represents the fraction of lithium nuclei that tunnel through the barrier and enter into a low energy reaction with the sulfur nuclei.

Of course, as I have mentioned in my articles, the Quantum Rabbit experiments are suggestive of low energy transmutation, and not yet definitive. Similar results to those achieved by Quantum Rabbit in laboratories around the world will be needed to establish low energy transmutation as a reality.

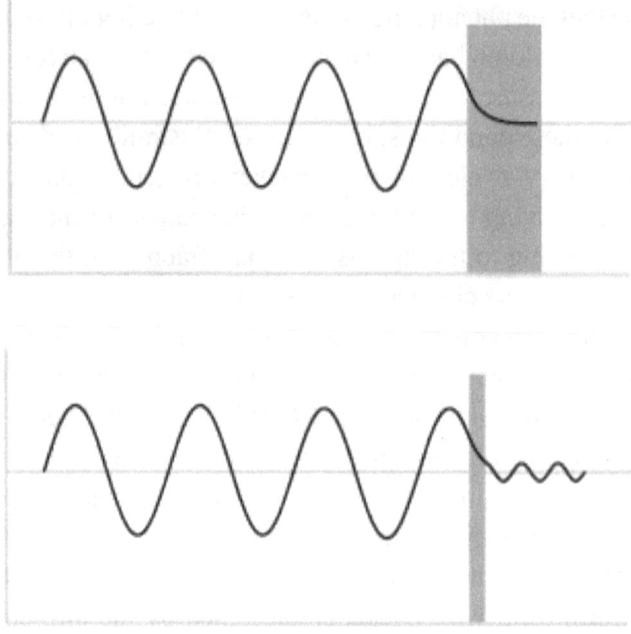

Source: Edward Esko, "Letter," *Infinite Energy* Issue 93, Sept./Oct. 2010.

13

THE POSSIBILITY OF PLUTONIUM REDUCTION

The issue of what to do with the nuclear waste left over from nuclear reactors and nuclear weapons programs remains a vexing problem in countries around the world. So far, no satisfactory solutions have emerged. Nuclear waste is clearly an impediment to a clean natural environment and poses a threat to world peace.

Environmental, health, and public interest groups including Greenpeace, Physicians for Social Responsibility, Friends of the Earth, and the Natural Resources Defense Council are united in opposition to the reprocessing of nuclear waste, a practice in which radioactive plutonium and uranium are separated from used or "spent" nuclear fuel from nuclear power reactors. The reprocessed waste is then reused as fuel.

In a May 2008 fact sheet entitled "Reprocessing: Dangerous, Dirty and Expensive–Why Extracting Plutonium from Nuclear Reactor Spent Fuel is a Bad Idea," the Union of Concerned Scientists stated that reprocessing would increase the risk of nuclear terrorism:

Less than 20 pounds of plutonium is needed to make a nuclear weapon. Commercial-scale reprocessing facilities handle so much of this material that it has proven impossible to keep track of it accurately. A U.S. reprocessing program would add to the worldwide stockpile of separated and vulnerable plutonium that sits in storage today, which totaled roughly 250 metric tons as of the end of 2005—enough for some 40,000 nuclear weapons. Reprocessing the U.S. spent fuel generated to date would increase this by more than 500 metric tons.

In a December 9, 2008 letter to then President-elect Obama, a coalition of 150 environmental and health groups pointed out that reprocessing would increase environmental contamination and threaten public health, cost hundreds of billions of dollars of taxpayer funds, and not solve the nuclear waste problem-not even in France: Although France reprocesses all its spent nuclear fuel, it is faced with the same difficulties the United States has in siting a permanent geologic repository. The proposed permanent repository site in Bure, France faces overwhelming public opposition, similar to Yucca Mountain in Nevada. In addition, reprocessing has polluted the environment, including the ocean as far away as the Artic Circle, and has created a stockpile of more than 80 metric tons of separated plutonium.

The coalition of public interest groups went on to urge the Obama administration to focus on securing nuclear waste at reactor sites. However well informed, this measure doesn't solve the problem of what to do with existing plutonium reserves. As we move forward, our goal should be not simply to bury radioactive waste in reactor sites, but to explore the possibility of reducing or even eliminating existing stocks of plutonium and other radioactive elements.

In *A Guide to the Elements Second Edition* (Oxford University Press) Albert Stwertka describes plutonium (Pu):

Plutonium [atomic number 94] is the most important of the transuranium elements, all of which follow uranium [atomic number 92] in the periodic table and all of which are artificially made.

In *Nature's Building Blocks: An A-Z Guide to the Elements* (Oxford University Press) John Emsley describes the destructive power of this man-made element, which has no role in nature or in the human body:

By the spring of 1945 several kilograms had been amassed, and the first atomic explosion, using plutonium, took place at Alamogordo, in the desert of New Mexico, on July 16. It detonated 6 kilograms of plutonium and was triggered by using small conventional explosives to force several pieces of plutonium together to give the critical mass necessary for a runaway chain reaction. The second plutonium explosion was in the form of a bomb, code-named 'Fat Man' which was dropped on the Japanese city of Nagasaki on 9 August 1945. The explosive capacity was equivalent to several thousand tons of TNT, killing about 70,000 citizens and wounding 100,000. The even more destructive hydrogen bombs are themselves triggered by a plutonium bomb whose explosion generates temperatures high enough to cause hydrogen atoms, in the form of the heavier isotopes, deuterium or tritium, to fuse together. Because of the fission-generated heat needed to generate the fusion reactions, these are classified as thermonuclear weapons. As hydrogen atoms fuse, they emit vastly more energy than an atomic bomb and give explosions equivalent to millions of tons of TNT.

In Quantum Conversion Theory, in which elements can be induced to change into each other, with relatively low inputs of energy, it may be possible to produce low energy nuclear fusion reactions, which in turn trigger simultaneous low energy nuclear fission reactions. It may be possible to prompt large, heavy, potentially unstable atoms, such as lead, uranium, or even plutonium, to shed protons and neutrons and convert into lighter more stable elements without destructive consequences. Quantum Rabbit low energy transmutation studies, conducted since 2005 and reported in *Infinite Energy*, suggest such a possibility. Holistic educator, Michio Kushi, suggested the possibility of low energy nuclear fusion/fission reactions as early as the 1960s.

The Quantum Conversion approach to the problem of plutonium reduction is simple, direct, and balanced. Atoms of sodium (atomic number 11) are subtracted from atoms of plutonium (atomic number 94) to form atoms of bismuth (atomic number 83). This low energy fission reaction is triggered by a low energy fusion reaction in which sodium-23 reacts with sulfur-34 (with oxygen as a fill gas under vacuum), to form cobalt-27. This reaction serves as a vector that "pulls" atoms of sodium from atoms of plutonium, thus yielding bismuth. Keep in mind that our proposal is to initiate these reactions at low temperature, pressure, and energy, without the release of radioactivity into the environment. On the contrary, our goal is to *reduce* the amount of radioactive material in the environment.

Proposals to store radioactive waste in salt mines are hinting at this reaction. There seems to be an intuitive awareness that salt, or sodium, can somehow neutralize radioactivity. Quantum Rabbit research with other large heavy nuclei suggests the reaction could be achieved in a sealed tube at approximately 3.5 torr with as little as a 4 amp discharge between anode and cathode. Large unstable atoms at the far end of the periodic table may actually be quite easy to fraction into two or more lighter elements. The formula is as follows:

$$Pu \rightarrow Na^* + Bi^* (+ \text{ neutrons})^*$$

*product of low energy fission

In addition to the reduction of plutonium, and the production of bismuth and cobalt, this reaction would most likely produce the discharge or release of neutrons. Plutonium has 19 known isotopes while bismuth has one natural isotope, 209Bi. The principal isotopes of plutonium, the way they may react to form bismuth, and the possible shedding of neutrons, is presented below:

$$^{238}Pu \rightarrow {}^{23}Na + {}^{209}Bi + {}^{6n*}$$
$$^{239}Pu \rightarrow {}^{23}Na + {}^{209}Bi + {}^{7n}$$
$$^{240}Pu \rightarrow {}^{23}Na + {}^{209}Bi + {}^{8n}$$
$$^{241}Pu \rightarrow {}^{23}Na + {}^{209}Bi + {}^{9n}$$
$$^{242}Pu \rightarrow {}^{23}Na + {}^{209}Bi + {}^{10n}$$

*n equals the number of neutrons discharged during the reaction

Quantum Rabbit is not equipped to handle plutonium or conduct the above-suggested research. However, we would be happy to serve as technical or theoretical advisors to established laboratories such as Oak Ridge, Los Alamos, Bhabha Atomic Research Center in India, and others around the world. As stated above, the goal of this research is to secure a clean natural environment for future generations and reduce potential nuclear threats to human health, peace, and wellbeing.

Source: Edward Esko "The Possibility of Plutonium Reduction," *Infinite Energy* Issue 84, March/April 2009.

14

LESSONS FROM JAPAN'S NUCLEAR CRISIS

The nuclear reactors in Fukushima, Japan, the site of multiple meltdowns
following the giant earthquake and tsunami on March 11, 2011

The earthquake and tsunami that occurred on March 11, 2011 have exposed
the weakness of Japan's nuclear program. Japan's policy on nuclear energy
is misguided and naïve. Prior to March 11, Japan's 54 reactors provided
some 30% of the country's total electricity production. There are ambitious

plans to build more reactors to increase this share to 41% by 2017 and 50% by 2030.

This policy is moving in the wrong direction. Rather than increasing its dependence on nuclear power to 50% by 2030, Japan should strive to reduce it's dependence by 50% by that date, with the stated goal of eliminating nuclear power before the end of the century, while, at the same time launching an all-out push to develop sustainable technologies such as wind, solar, hydro, geothermal and biothermal.

Simply substituting green technologies for nuclear power will not solve the whole problem, however. There remains the problem of what to do with spent nuclear fuel, which, as many have discovered in this latest crisis, is highly radioactive and highly dangerous.

The worldwide inventory of spent nuclear fuel was 220,000 tons in the year 2000, and is increasing by about 10,000 tons a year. Even with billions of dollars spent on a variety of disposal options, the nuclear industry and governments have not come up with a feasible and sustainable solution.

Most current proposals for dealing with nuclear waste involve burying it in deep underground sites. Whether the storage containers, the store itself, or the surrounding rocks will offer enough protection to stop radioactivity from escaping in the long-term remains unknown.

Japan is currently pursuing a policy of reprocessing nuclear waste at a cost of trillions of yen. Recycling nuclear waste is a distortion of the concept of recycling. Used nuclear fuel has been shipped to the UK and France where it has been reprocessed and returned to Japan as reactor fuel, notably as mixed oxide (MOX) fuel. Labeling this practice "recycling" is a deception designed to convey a "green" veneer to the process.

Rather than "recycling" nuclear waste, Japan should seek to eliminate it, beginning with highly toxic plutonium. The dark history of this destructive manmade element is especially relevant to Japan. The first atomic bomb tested at Alamogordo in 1945 used plutonium to create critical mass necessary for an atomic explosion. This process was repeated on August 9, 1945 with the explosion of a plutonium bomb over the city of Nagasaki.

Rather than investing trillions of yen in additional reprocessing facilities, and essentially not dealing with the problem, those funds should be invested in promising new technologies that offer the possibility of eliminating these toxic materials.

Promising new technologies include using microorganisms to convert toxic waste into harmless elements through the process of biological transmutation. Another promising technology is the laboratory process of cool fusion and cool fission, in which heavy unstable radioactive nuclei are induced to shed mass and convert to more stable non-radioactive elements at low temperatures, pressures, and energies. In this process, suggested by Quantum Rabbit research, plutonium-239 and uranium-235 are converted to bismuth-209 and lead-208 which, although not ideal as end points in the nuclear reduction process, are nevertheless less dangerous than their radioactive counterparts (see "The Possibility of Plutonium Reduction," *IE* Issue 84, March/April 2009.)

We propose that the nuclear industry take the lead in research ways to eliminate nuclear waste. A global "clean the planet" campaign funded by the nuclear industry would help solve the problem. Even If the proposed conversion to non-toxic materials reduces the problem of nuclear waste, because of it's inherent dangers, nuclear power should nevertheless take a back seat to the development of sustainable technologies. Given the events of March 11, a global moratorium on construction of any new nuclear power plants seems like a prudent course of action coupled with a plan to implement nuclear phase-out as green replacement technologies are phased in.

The events of March 11 have brought our choices into sharp focus. Japan and the rest of the world have the choice of whether to continue on the current path, and face the possibility of future accidents and disasters, or to create a safer path toward genuine sustainability.

Source: Edward Esko, "Letter," *Infinite Energy* Issue 98, July/August 2011.

Part III

Corking the Nuclear Genie

The problem is how to keep radioactive waste in storage
until it decays after hundreds of thousands of years. The
geologic deposit must be absolutely reliable as the quantities
of poison are tremendous. It is very difficult to satisfy
these requirements for the simple reason that we have had
no practical experience with such a long-term project.
HANNES ALFVEN, NOBEL LAUREATE IN PHYSICS

Nuclear remediation is a critical problem, which I'm sure can
be solved in a relatively low energy plasma environment. So I
will be pleased to refer to your work appropriately in future.
WAL THORNHILL, ELECTRIC UNIVERSE

The giant casks of spent nuclear fuel loom over the future
like a modern, highly toxic Stonehenge. They bear silent
witness to our inability, at least for now, to solve the problem
of nuclear waste. They are the sentinels of our ignorance.
EDWARD ESKO

FOREWORD

What if cesium-137, strontium-90, uranium-235, and other highly radioactive elements could be converted into harmless, stable isotopes? Radioactive cesium, the main byproduct of nuclear accidents in Chernobyl and Fukushima, has a half-life of 30 years and will pose a serious health risk to people for about 300 years. (As a rule of thumb, a radioactive substance is hazardous for ten times its half-life, or enough time to reduce its lethality to about 0.1%.) Two of the most long-lived fission products are technetium-99 (half life 220,000 years) and iodine-129 (half life 15.7 million years). Other lethal isotopes are neptunium-237 (half life 2 million years) and plutonium-239 (half life 24,000 years). One of the most long-lived, uranium-235 has a half-life of 700 million years and, applying the rule of ten, will not completely decay for 7 billion years. That is the estimated time the sun will last and hence all future life on earth. Fortunately, its lethality evaporates long before that time, as do other Methuselah-like fission products whose half-lives extend billions of years!

The Cold War and the era of nuclear power have resulted in millions of tons of nuclear waste for which there is no permanent storage. All current solutions, including burying waste in salt mines, are inherently unstable and will inevitably contaminate the planet. But what if, rather than trying to contain or shield these deadly isotopes, radioactive cesium-137 could be converted into the highly sought-after rare earth element neodymium, strontium-90 into ruthenium, the first element in the scarce platinum group of elements, or highly radioactive iodine-129 into a stable and utterly harmless isotope of barium? Or in the case of uranium, which has no nonradioactive form, transformed into bismuth, the last stable element in the Periodic Table? At a stroke, the nuclear genie could be put back in the bottle and useful raw materials produced.

The transmutation of elements, or Cool Fusion, is the goal of Quantum Rabbit (QR), the small macrobiotic company I helped launch with Edward Esko and Woody Johnson. In experiments at laboratories in New England over the last eight years, QR has successfully created small amounts of about 20 key industrial metals, including iron, copper, titanium, silver, gold, aluminum, and scandium, from carbon, silicon, and other common elements. The tests build on pioneer studies conducted by Louis Kervran, a French biochemist, and macrobiotic educators George Ohsawa and Michio Kushi in the 1960s.

In early 2013, QR conducted its first isotope studies. In vacuum tube experiments in Owl's Head, Maine, we set out to transmute iodine-127 into barium-134 through low energy nuclear reactions (LENR). Lithium-7 was used as a catalyst, producing the formula: iodine-127 + lithium-7 → barium-134. Barium, a stable element in the center of the Periodic Table, is best known as a contrast for radioactive X-rays in the barium enema test. Barium was also in the news this spring when the ratio of barium to calcium in teeth was used to determine how long Neanderthals nursed their children (7 months). As this archeology study demonstrated, the barium remained stable for over 100,000 years.

The QR study involved three separate tests. Upper and lower copper electrodes were inserted into both ends of a vertical vacuum tube. A lithium plug was then inserted in the center of the lower electrodes. Pure iodine crystals were placed on top the lithium inserts. In a separate test, a small piece of lithium was placed in the center of the lower copper electrode surrounded by iodine crystals instead of the plug.

After pumping down to vacuum, oxygen was admitted and the power turned on. When the arc was established and plasma struck, additional heat was provided by a hand-held torch. The arc was maintained for about 10-15 minutes, at which time the power was disconnected and the tubes allowed to cool. Residue from each test was collected and sent to Northern Analytical Laboratory in Londonderry, N.H. for independent analysis. As predicted, barium appeared in all samples, ranging from 3.5 ppm to 1.8 ppm, to 463 ppm. The distribution of barium isotopes in the test samples significantly varied from the distribution in nature, suggesting the possibility that detected barium was newly created through low energy transmutation and not from contamination in the laboratory. Barium-134 occurs in only 2.4% of all the barium on the planet. In the QR tests, it measured up to 4.04% or nearly double. Barium-132, which is found at 0.101% in nature appeared up to 3.92%, or about 40 times higher than normal.

Northern Analytical Laboratory was so startled by the anomalous findings that it reevaluated the tests on its own initiative. It confirmed the results but theorized that the spectroscopic analysis of barium overlapped with that of zinc, which has a similar profile, accounting for the discrepancies. Since there was no zinc used in the experiments, however, the chance of contamination was remote. Most likely, low energy fusion between iodine (atomic number 53) and lithium (atomic number 3) could explain the consistent appearance of barium (atomic number 56) in all tests. The distribution of barium isotopes can be accounted for by the fusion of iodine-127 and lithium-7. That formula was predicted beforehand. Barium-132 may have arisen after the ejection of a neutron at the moment of fusion between the lithium and iodine.

Such studies need to be replicated and eventually radioactive materials introduced. For that, a university or government nuclear laboratory would need to be involved. And then when proof of concept is definitely established, engineers would need to be brought in to scale the process up for practical use, such as designing on site remediation units that could immediately decontaminate nuclear waste from atomic power plants, factories, and hospitals. A coordinated effort by small companies like QR, NGOs, industry, the military, and government could creatively solve the nuclear storage crisis within our lifetimes, bequeath a nuclear-free planet to our children and future generations, and cork the nuclear genie once and for all.

Alex Jack
The Berkshires
October 26, 2013

Introduction

Once each season for several years the cool fusion research team would meet at the Moore Mill building in Bellows Falls, Vermont, to conduct tabletop carbon-arc experiments. Alex Jack and I would drive up from Western Massachusetts. The drive took us east on the Mass Pike and then north up route 91 past Brattleboro to the Rockingham exit. Bellows Falls is on the Connecticut River. Thirty miles downriver, in the town of Vernon, is the Vermont Yankee nuclear power plant.

Vermont Yankee epitomizes the chaotic state of affairs within the nuclear industry. The amount of radioactive waste stored at Vermont Yankee is substantial. It exceeds that of all four damaged reactors at Fukushima in Japan. The used fuel rods at Vermont Yankee are about one million times more radioactive than they were before being used in the reactor. The rods are hot enough to catch fire if they are not stored under water. Five hundred tons of spent fuel is now being stored in pools of water seven stories above ground. [1] Following decades of conflict with local residents and with the State of Vermont, it was announced that Vermont Yankee would close by the end of 2014. The Yankee plant went on line in 1972 and is the same General Electric boiling water system as the failed nuclear reactors at Fukushima. Following the disaster at Fukushima, Entergy, the owner of Vermont Yankee, was being pressured to make expensive modifications to improve safety.

According to nuclear engineer Arnie Gunderson, "The problem is Fukushima modifications are coming due, that's close to $100 million, plus the leaky condenser, that's another $100 million. So it just made no economic sense." [2] Financial experts point to low natural gas prices as another factor in the closure of the plant, explaining that competition from cheap natural gas-fired electric plants severely limited Vermont Yankee's margins. Julien Dumoulien-Smith, an energy analyst with UBS financial services predicted, "This is likely a bellwether as far as it goes for the nuclear industry. In some sense, this is the first or perhaps the second of a large wave of potential nuclear power plant retirements in the country." [3]

The closing of the plant means that Vermont Yankee will no longer be producing radioactive waste. However, still unresolved is what will happen with the tons of highly radioactive spent fuel rods stored at the site. Local residents are justifiably concerned. One resident, who is the director of a local citizen advocacy group, stated, "There are at least 530 tons of high-level (radioactive) waste, which we've said still needs to come out of that spent fuel pool. This isn't over. The struggle is now about cleanup." [4]

Another resident stated, "The banks of the river, within a 500-year flood plain, is not the best place to store high-level radioactive waste for even a short period of time. Unfortunately, there is no long-term storage in the U.S., and we're probably stuck with it there, just like at Yankee Rowe." [5] Yankee Rowe is the former nuclear plant in nearby Rowe, Massachusetts that was shut down in 1992. High-level radioactive waste from the plant is now in temporary storage in 16 dry casks at the site. The giant casks of spent nuclear fuel loom over the future like a modern, highly toxic Stonehenge. They bear silent witness to our inability, at least for now, to solve the problem of nuclear waste. They are the sentinels of our ignorance.

Dry casks at Yankee Rowe

A third resident expressed concern about the safety of Vermont Yankee's phase-out: "Now until the fall of 2014 will be the most dangerous year of their operation of the plant, because the plant will be older than ever, parts will be more brittle than ever, they will be more reluctant than ever to repair and replace parts ... and workers who have already been let go are going to be leaving in much larger numbers. To expect that Entergy is going to be taking every single precaution to keep the plant as safe as it possibly can be is unfortunately unrealistic." [6] A member of the citizen advocacy group, the New England Coalition, explained the situation at Vermont Yankee quite succinctly: "One fundamental purpose of our advocacy has always been to try to protect the public and the environment from nuclear waste—waste in the fuel, in the reactor, in the pool, out-in-the-yard, soon to be released in the next reactor or fuel handling accident, and out on the wind. Soon, Entergy Vermont Yankee, a nuclear waste pile that generated electricity will stop generating electricity—and it will either be mothballed or promptly torn apart, but it will be, absent electricity generation, just a nuclear waste pile ... from which the public and the environment need to be protected."[7]

On a planetary scale, there are more than 430 locations around the world where nuclear waste continues to accumulate. Most is stored at individual reactor sites. Nuclear reactors on planet earth create about 10,000 metric tons of spent nuclear fuel each year. Thus, the problem gets worse with each passing day. The shutdown and decommission of nuclear power plants solves only the problem of new nuclear waste. It does nothing to solve the problem of already existing waste. Moreover, the process is expensive (between $300 million to $5.6 billion per unit), time-consuming, and hazardous to workers and the natural environment. It opens a window for disaster caused by human error, accident, or sabotage.

In the U.S. there are 13 reactors that have shut down and are in the process of decommissioning. None have fully completed the process that can last as long as 100 years. The timeframes when dealing with nuclear waste are enormous; they range from 10,000 years to millions of years. Storage of nuclear waste, whether "temporary" or "permanent," does not solve the problem. It simply passes it on to future generations. The very existence of nuclear waste is itself the problem. For the sake of future generations, we need to seriously investigate promising ideas not just for storing nuclear waste, but also for actually *getting rid* of it.

Toward that end, Quantum Rabbit LLC has been conducting table-top research on the low energy transmutation. Transmutation is defined as transforming one atom into another by changing its nuclear structure. Low energy transmutation attempts to achieve this with simple tabletop equipment, using electric power from car batteries, solar panels, a HUBERT® portable generator, and the wall socket, with relatively low temperatures and low pressures created in simple glass vacuum tubes. This is in contrast to the high-energy accelerator transmutation of waste (ATW) being studied around the world as a possible solution to the radioactive waste problem.

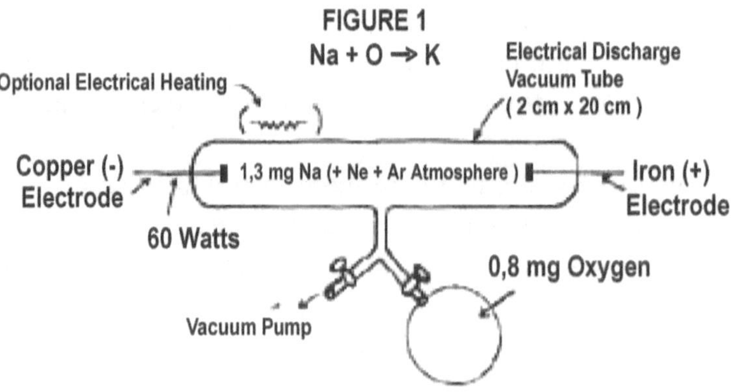

The Ohsawa sodium into potassium experiment

The Quantum Rabbit experiments are based on the work of George Ohsawa, the Japanese philosopher and educator, who, together with French biochemist Louis Kervran, developed the theory of biological and elemental transmutation. In 1964, Ohsawa claimed to have transmuted sodium into potassium in a tabletop experiment conducted in Tokyo. [8] Ohsawa claimed that this transmutation took place through a process of low energy fusion in which two lighter elements, sodium and oxygen, fused to form the heavier element potassium. A diagram of his experiment is shown above. Sodium has the atomic number 11 and the atomic weight 23. Oxygen has the atomic number 8 and the atomic weight 16. If we add the atomic number and weight of sodium with the atomic number and weight of oxygen, we get the atomic number (19) and weight (39) of potassium. The formula is written as follows, with the plus sign (+) indicating a fusion reaction and the arrow ($\rightarrow$) indicating the fusion product:

$$^{23}_{11}Na + {}^{16}_{8}O \rightarrow {}^{39}_{19}K$$

Following this initial experiment, Ohsawa began researching the transmutation of carbon into iron, with the addition of oxygen. In Cambridge, Massachusetts, Michio Kushi, a student of Ohsawa, performed experiments in which carbon and oxygen were apparently fused to form iron. [9] The year was 1965. The carbon into iron experiment was done in the open air and is another example of low energy fusion. Graphite (carbon) powder was placed on a copper plate. A graphite rod was connected to several car batteries. An arc was struck when the rod was placed near the graphite powder. The arc sucked in oxygen and nitrogen from the surrounding air. The process was repeated and the remaining powder was found to be magnetic and to contain trace amounts of iron. The low energy fusion formula can be written as follows:

$$_2(^{12}_{6}C + {}^{16}_{8}O) \rightarrow {}^{56}_{26}Fe + {}^{2\ protons}$$

Forty years later, in 2004, I joined with Alex Jack and Woody Johnson to form Quantum Rabbit LLC (QR), a Massachusetts company, to develop the ideas and experiments of Ohsawa and Kushi. We established a small lab in Nashua, New Hampshire to begin experiments on low energy transmutation using custom designed glass vacuum tubes. Carbon arc experiments were also started at Woodland Energy in Bellows Falls, Vermont. In 2009, the vacuum lab moved to Owls Head, Maine.

Table 1. Results of Carbon Arc Experiments

Element	Concentration (ppm*)
Silicon	10,500
Magnesium	1800
Iron	4700
Aluminum	7800
Titanium	440
Scandium	35
Cobalt	160
Nickel	1120

*Parts per million

As with the experiments conducted forty years earlier, in our experiments, when subjected to the arcing process, carbon displayed magnetic activity. Placing the treated graphite powder on white paper and waving a neodymium magnet underneath it repeatedly confirmed magnetic properties. [10] Not only was the graphite magnetic, but also when analyzed independently, it showed traces of the elements shown in Table 1. [11] Low energy transmutation may help explain the presence of anomalous elements in treated graphite (Table 2).

Table 2. Possible Low Energy Fusion Reactions in Treated Graphite*

Magnesium-24	Calcium-40
$^{12}C + {}^{12}C \rightarrow {}^{24}Mg$	$^{24}Mg + {}^{16}O \rightarrow {}^{40}Ca$
Aluminum-27	Titanium-44
$^{12}C + {}^{15}N \rightarrow {}^{27}Al$	$^{28}Si + {}^{16}O \rightarrow {}^{44}Ti$
Silicon-28	Iron-56
$^{12}C + {}^{16}O \rightarrow {}^{28}Si$	$_2(^{12}C + {}^{16}O) \rightarrow {}^{56}Fe\ ^{+2\ protons}$

Potassium-39   Scandium-45
$^{26}Mg + {}^{14}N \rightarrow {}^{39}K$ $^{30}Si + {}^{15}N \rightarrow {}^{45}Sc$

*The gases involved in these reactions, oxygen (O) and nitrogen (N), are from the atmosphere.
Elements shown with their atomic weights

Following a vacuum arc experiment in Nashua on May 30, 2008, 1,500-ppm copper was unexpectedly detected on an electrode made of stainless steel. Aside from stainless electrodes, lithium and oxygen were the only elements introduced into the experiment, suggesting the possibility that the copper was newly created through transmutation. The stainless electrodes were composed mostly of iron (Fe). Low energy fusion of the iron in the electrodes with the lithium (Li) test material could explain the appearance of copper:

$$^{56}Fe + {}^{7}Li \rightarrow {}^{63}Cu$$

The test was repeated on Dec. 30, 2008 with similar results. In the Dec. 30 test, copper was found on the electrodes at 557 ppm. [12] These results appeared to confirm the theory of low energy transmutation and prompted Quantum Rabbit to initiate a series of metal vapor tests using a variety of electrode elements and catalyst materials. The metal vapor experiments are described in our book, *Cool Fusion* (Amber Waves, 2012). The results of the metal vapor experiments are summarized in Table 3.

Table 3. Results of Vacuum Arc Experiments

Element	Concentration*	Catalyst	Electrode
Copper	1,500	Lithium	Stainless
Germanium	3,570	Lithium	Copper
Tin	5	Lithium	Silver
Potassium	750	Lithium/sulfur	Copper
Palladium	181	Sulfur	Zinc

Element	Concentration*	Catalyst	Electrode
Strontium	3.5	Oxygen	Zinc
Chromium	10	Sulfur/oxygen	Copper
Aluminum	342	Boron/oxygen	Copper
Scandium	18	Boron/sulfur	Copper
Selenium	12	Boron	Copper

*Parts per million

In every case, the results were predicted beforehand. Formulas based on the theory of low energy transmutation guided the protocol for each experiment. Test samples were carefully collected and sent to an outside lab for analysis by ICP (Inductively Coupled Plasma Spectroscopy). Low energy fusion formulas explaining the vacuum arc anomalies are presented in Table 4.

Table 4. Possible Low Energy Fusion Reactions in Vacuum Arc Experiments*

Copper-63

$^{56}Fe + {}^{7}Li \rightarrow {}^{63}Cu$

Strontium-86

$^{68}Zn + {}^{18}O \rightarrow {}^{86}Sr$

Germanium-72

$^{65}Cu + {}^{7}Li \rightarrow {}^{72}Ge$

Chromium-50

$^{34}S + {}^{16}O \rightarrow {}^{50}Cr$

Tin-116

$^{109}Ag + {}^{7}Li \rightarrow {}^{116}Sn$

Aluminum-27

$^{16}O + {}^{11}B \rightarrow {}^{27}Al$

Potassium-39

$^{32}S + {}^{7}Li \rightarrow {}^{39}K$

Scandium-45

$^{34}S + {}^{11}B \rightarrow {}^{45}Sc$

Palladium-102

$^{68}Zn + {}^{34}S \rightarrow {}^{102}Pd$

Selenium-76

$^{65}Cu + {}^{11}B \rightarrow {}^{76}Se$

*Elements shown with their atomic weights

In 2005 I began working on formulas for the low energy fission of heavy elements. Low energy fission is the opposite of low energy fusion. In low energy fission, a heavy element, such as lead or bismuth, splits into two lighter elements. In my research, I noted textural and energetic similarities between the elements lead, gold, and lithium. I developed the following formula based on these observations:

$$^{204}Pb \rightarrow {}^{197}Au + {}^{7}Li$$

The first sketch I did of the experiment is shown below. In it I proposed initiating the low energy fission of lead into gold by a low energy fusion reaction in which lithium fuses with oxygen to form sodium. The key to achieving this transmutation is to pull atoms of lithium out of atoms of lead. Michio Kushi refers to this process as "nuclear subtraction." As he states in *The Philosopher's Stone*, "After studying this problem, I began to see several possible means for transmuting gold using the addition method. But unless I could understand the subtraction method, my understanding would have remained incomplete." [13]

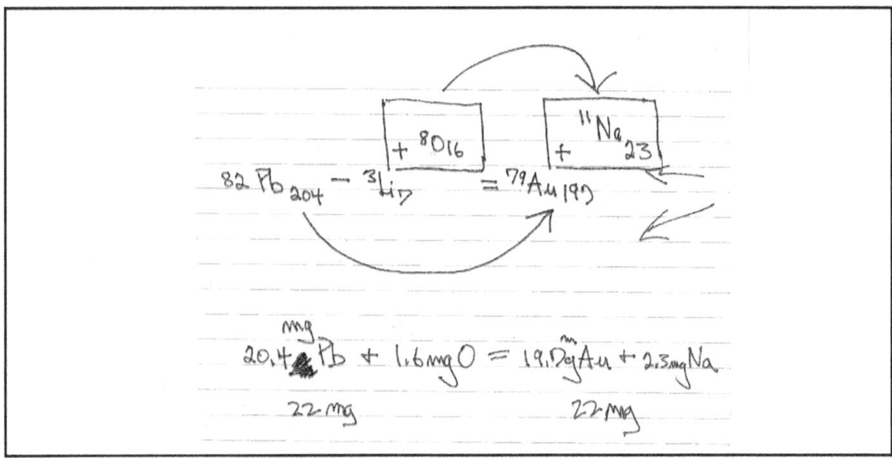

July 2005 sketch of the Pb $\rightarrow$ Au experiment

After doing experiments with lithium, sodium, oxygen, and sulfur, I started to think that it would be easier for lithium to react with solid sulfur rather than with gaseous oxygen. In the theory of low energy transmutation, sulfur is a form of crystalized oxygen. It is formed by the fusion of two oxygen atoms ($^{16}O + {}^{16}O \rightarrow {}^{32}S$). So when it became time to do the actual experiment, we decided to go with sulfur, with the goal being to trigger the low energy fission of lead into gold through the following reaction:

$$^{7}Li + {}^{32}S \rightarrow {}^{39}K$$

Lithium-7 + sulfur-32 $\rightarrow$ potassium-39

I hypothesized that the low energy fusion of lithium and sulfur would release the energy needed to fission lead into the two lighter elements. That is why, in each of the experiments performed to test this hypothesis, in addition to gold, our analysis tested for potassium. The presence of potassium could provide evidence that at least one of the predicted low energy nuclear reactions took place. The QR team tested the Pb $\rightarrow$ Au formula over a four-year period, beginning in 2008.

The success of these experiments led me to speculate about the possibility of using low energy methods to transmute the radioactive isotopes found in nuclear waste into stable non-radioactive elements. In theory, the process of subtracting lithium or another light element from super-heavy elements could be used to condense the decay cycle of radioactive elements such as uranium-235 and plutonium-239 from thousands or millions of years to a few years at most. Proposed experiments are outlined in my paper, "Low Energy Transmutation of Nuclear Waste."

On a parallel track, low energy fusion could also be deployed to instantly convert radioactive fission products like iodine-129, technetium-99, and cesium-137 into useful elements like barium, palladium, and neodymium. Proposals for transmutation of these radioactive waste products are also presented in my paper. On April 10, 2013, QR conducted the first of what is hoped will be a series of preliminary tests on the remediation of nuclear waste. The test involved the transmutation of stable iodine-127 into barium-134 through a simple low energy fusion reaction. Radioactive iodine-129 is created by the fission of uranium and plutonium in nuclear reactors. With a half-life of 15.7 million years, iodine-129 poses a significant threat to human health and the environment. Successful transmutation of stable iodine-127 could provide a pathway for research on converting radioactive iodine-129 into non-radioactive barium. The 2013 experiment tested the formula:

$$^{127}I + {^7}Li \rightarrow {^{134}}Ba$$

As predicted, barium appeared in all test samples, ranging as high as 463 ppm from a starting concentration (in the iodine used in the test) of 0.43 ppm, more than a thousand-fold increase. Moreover, the distribution of isotopes in the treated sample varied considerably from the distribution of isotopes found in nature. Novel isotope distribution is considered the gold standard for ruling out contamination and proving transmutation. The experiments are presented in detail in my paper, "Appearance of Barium in Lithium-Iodine Plasma."

Clearly, much work needs to be done. We plan to continue experiments, but with limited time and resources, our work remains preliminary, like that of Steve Jobs and Steve Wozniak in their garage in Silicon Valley or the Wright Brothers in their bicycle shop in Ohio. Our hope is to eventually partner with established labs or universities to take our preliminary work to the next level.

Nuclear energy was unleashed seventy years ago with the Manhattan Project. In today's dollars, the creation of nuclear power cost about $26-billion. With research and development in 30 locations, the project employed 130,000 people. It may require a similar effort today, with researchers around the globe committed to the same goal, to put the nuclear genie back into the bottle. We owe it to future generations to at least try.

Edward Esko
Pittsfield and Lenox, Mass.
Autumn Equinox, 2013

NOTES:

[1] Vermont Citizens Action Network, "Where does Vermont Yankee's waste go?" vtcitizens.org/waste.

[2] Dillon, John, "Citing Economics, Entergy to Close Vermont Yankee by End of 2014," digitial.vpr.net, August 27, 2013.

[3] Ibid.

[4] "Anti-nuclear community celebrates Vermont Yankee closing," times-argus.com, September 2, 2013.

[5] Ibid.

[6] Ibid.

[7] Ibid.

[8] Kushi, Michio, *The Philosopher's Stone*, One Peaceful World Press, Becket, Mass., USA, pp. 23-28, 1994.

[9] Kushi, Michio, *The Philosopher's Stone*, One Peaceful World Press, Becket, Mass., USA, pp. 28-33, 1994.

[10] Esko, Edward and Jack, Alex, *Cool Fusion*, second edition, Amber Waves, Becket, Mass. USA, pp. 56-60, 2012.

[11] Ibid.

[12] Esko, Edward and Jack, Alex, *Cool Fusion*, second edition, Amber Waves, Becket, Mass. USA, pp. 65-70, 2012.

[13] Kushi, Michio, *The Philosopher's Stone*, One Peaceful World Press, Becket, Mass., USA, pp. 33-36, 1994.

1

Low Energy Transmutation of Nuclear Waste

Abstract

Quantum Rabbit (QR) research on the low energy fusion and fission (low energy nuclear reactions, or LENR) of various elements indicates possible pathways for applying that process to reducing nuclear materials. In a New Energy Foundation (NEF)-funded test conducted at Quantum Rabbit lab in Owls Head, ME, QR researchers initiated a possible low energy fission reaction in which lead-204 fissioned into lithium and gold ($^{204}Pb \rightarrow {}^{7}Li + {}^{197}Au$). [1] This reaction may have been triggered by a low energy fusion reaction in which lithium fused with sulfur to form potassium ($^{7}Li + {}^{32}S \rightarrow {}^{19}K$). These results confirmed earlier findings showing apparent low energy fusion and fission reactions. [2]

Moreover, subsequent research with boron indicates apparent low energy fusion reactions in which boron fuses with oxygen to form aluminum and with sulfur to form scandium. [3] At the same time, the QR group has achieved what appear to be low energy transmutations of carbon using carbon-arc under vacuum and in open air. [4] The research group at QR believes these processes can be adapted to accelerate the natural decay cycle of uranium-235, plutonium-239, radium-226, and the fission products cesium-137, iodine-129, and technetium-99, with the long-term potential of reducing the threat posed by radioactive isotopes to human health and the environment.

URANIUM-235

Radioactive uranium is the primary constituent of spent nuclear fuel. The half-life of uranium-235 is more than 700 million years. The first step in this process, the alpha decay of uranium-235 into thorium-231 consumes the bulk of this enormous span. The half-lives of the isotopes that follow thorium-231 total approximately 33,000 years with the stable isotope lead-208 as the conclusion of the process.

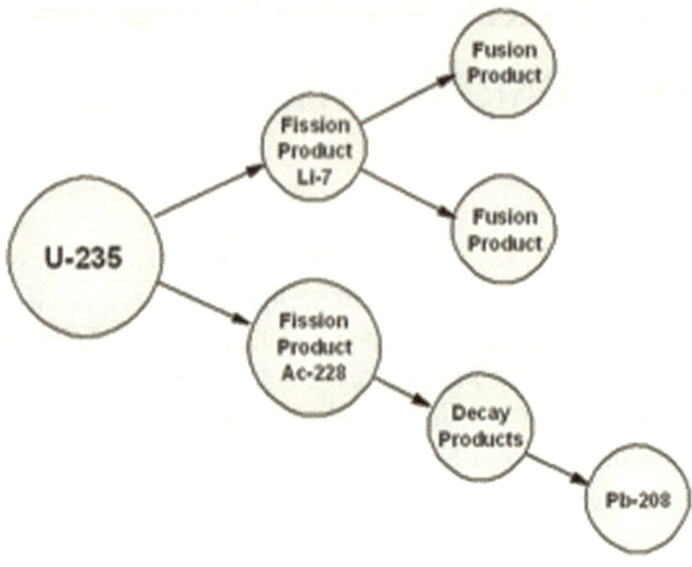

Low energy transmutation of uranium-235

The QR research indicates it may be possible to intervene in the decay cycle of uranium in order to reduce the amount of time needed to achieve its transmutation into lead. The most obvious window for intervention is at the beginning of the cycle, by inducing uranium-235 to fission into one of the lighter isotopes in the radioactive decay chain. We propose using lithium, the catalyst element in the studies cited above, as the catalyst for the following low energy fission reaction:

$$^{235}U \rightarrow {}^{7}Li + {}^{228}Ac$$

Uranium-235 → lithium-7 + actinium-228

According to this hypothesis, the low energy fusion of lithium with sulfur, resulting in potassium, triggers the low energy fission of uranium into lithium and actinium. The low energy fusion reaction can be written as follows:

$$^7Li + {}^{32}S \rightarrow {}^{39}K$$

Lithium-7 + sulfur-32 → potassium-39

These reactions are summarized above. If achieved, they set in motion the natural decay cycle beginning with actinium-228 and ending with lead-208 as shown below. Note that the low energy nuclear reaction (LENR) that causes the uranium-235 to fission into actinium-228 results in U-235 being cycled downstream into the natural decay chain of thorium-232. [5]

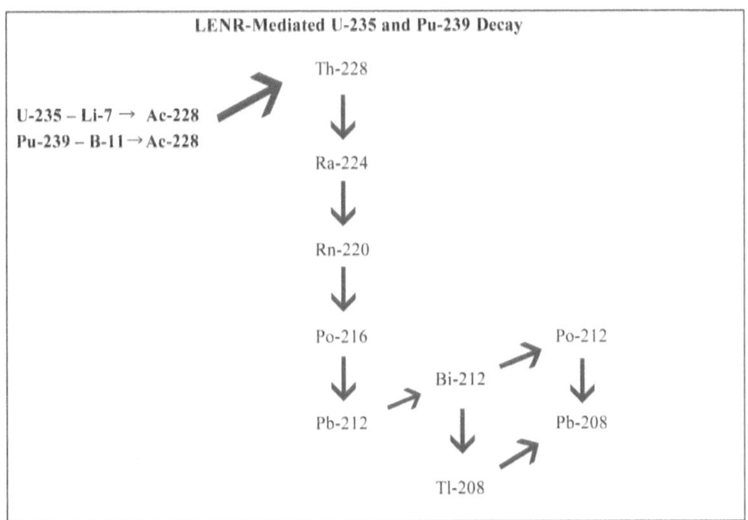

Accelerated Decay Series: U-235 and Pu-239. Downward arrows represent alpha decay; upward arrows beta decay.

If actinium-228 is produced as predicted, and the natural decay-cycle indicated above is set in motion, the half-life of uranium-235 is compressed from over 700-million years to slightly over 1.9 years. The process is summarized in the formula:

$$^{235}U \rightarrow {}^{7}Li + {}^{228}Ac \rightarrow {}^{208}Pb$$

Uranium-235 → lithium-7 + actinium-228 (thorium-232 decay cycle) → lead-208

Plutonium-239

There is a significant quantity of deadly plutonium-239 in spent nuclear fuel. Plutonium-239 has a half-life of 24,000 years. The QR research group achieved promising results with the low energy fusion of boron. A series of experiments for the possible reduction of plutonium-239 similar to the QR boron experiments can be designed using boron as the catalyst element. The low energy fission reaction we propose testing is as follows:

$$^{239}Pu \rightarrow {}^{11}B + {}^{228}Ac$$

Plutonium-239 → boron-11 + actinium-228

In theory, this low energy fission reaction can be triggered by several low energy fusion reactions: [6]

$$^{11}B + {}^{16}O \rightarrow {}^{27}Al$$

Boron-11 + oxygen-16 → aluminum-27

$$^{11}B + {}^{34}S \rightarrow {}^{45}Sc$$

Boron-11 + sulfur-34 → scandium-45

Once again, if these low energy transmutations are successful in producing actinium-228, like U-235 in the formula described previously, Pu-239 would be cycled downstream into the thorium-232 decay chain with the end product being the stable isotope lead-208. This process can be summarized as follows:

$$^{239}Pu \rightarrow {}^{11}B + {}^{228}Ac \rightarrow {}^{208}Pb$$

Plutonium-239 → boron-11 + actinium-228 (thorium-232 decay cycle) → lead-208

Radium-226

Contamination by radium-226 continues to be a problem at U.S. military installations and other sites around the world. Radium-226 is part of the U-238 decay chain with a half-life of 1,600 years. With low energy transmutation, it may be possible to compress this time frame considerably by achieving the low energy fission of Ra-226. QR research on carbon-arc may offer a method for achieving this possibility. Numerous low energy transmutations have been reported, both in open air and under vacuum. [7] These low energy fusion reactions could possibly be used to prompt the low energy fission of radium-226, compressing the half-life of radium and accelerating the natural decay cycle from more than 1,600 years to approximately 22 years. The low energy fission reaction we propose testing is as follows:

$$^{226}Ra \rightarrow {}^{12}C + {}^{214}Pb$$
Radium-226 $\rightarrow$ carbon-12 + lead-214

This low energy fission reaction could possibly be triggered by low energy fusion reactions such as those between carbon and oxygen noted in QR carbon-arc research:

$$^{12}C + {}^{12}C \rightarrow {}^{24}Mg$$
Carbon-12 + carbon-12 $\rightarrow$ magnesium-24

$$^{12}C + {}^{16}O \rightarrow {}^{28}Si$$
Carbon-12 + oxygen-16 $\rightarrow$ silicon-28

$$^{12}C + 2({}^{16}O) \rightarrow {}^{44}Ti$$
Carbon-12 + 2(oxgyen-16) $\rightarrow$ titanium-44

$$^{12}C + {}^{32}S \rightarrow {}^{44}Ti$$
Carbon-12 + sulfur-32 $\rightarrow$ titanium-44

$$_2({}^{12}C + {}^{16}O) \rightarrow {}^{56}Fe \text{ (+2 protons)}$$
$_2$(Carbon-12 + oxygen-16) $\rightarrow$ iron-56 + two protons

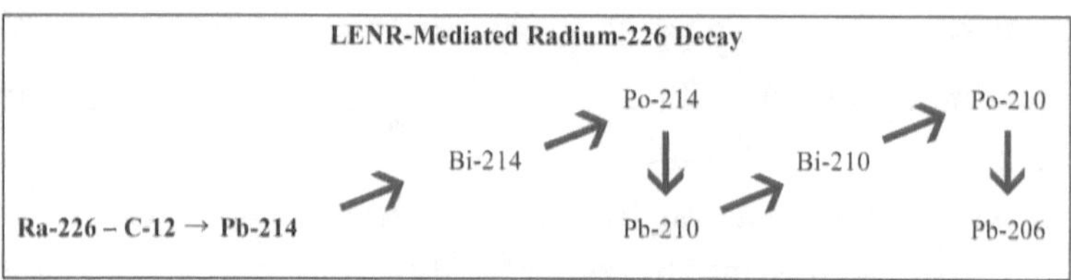

Accelerated Decay Series: Ra-226. Downward arrows represent alpha decay; upward arrows beta decay.

Cesium-137

Cesium-137, a product of nuclear fission is a major radionuclide in spent nuclear fuel. It is of major concern for Department of Energy environmental management sites and has a half-life of 30 years. It decays by emitting a beta particle. Its decay product, barium-137m (the "m" is for metastable) stabilizes by emitting an energetic gamma ray with a half-life of approximately 2.6 minutes. It is this decay product that qualifies cesium-137 as a radiation hazard.

The environmental dangers posed by cesium-137 were highlighted by the crisis at Fukushima Daichi reactor in Japan. Writing in the Proceedings of the National Academy of Sciences, [8] an international team of scientists described the threat posed by cesium-137:

The largest concern on the cesium-137 (^{137}Cs) deposition and its soil contamination due to the emission from the Fukushima Daiichi Nuclear Power Plant (NPP) showed up after a massive quake on March 11, 2011. Cesium-137 (^{137}Cs) with a half-life of 30.1 y causes the largest concerns because of its deleterious effect on agriculture and stock farming, and, thus, human life for decades. Removal of ^{137}Cs contaminated soils or land use limitations in areas where removal is not possible is, therefore, an urgent issue."

Contamination by cesium-137 was a major problem following the Chernobyl disaster. As John Emsley states: [9]

Uranium fuel rods in nuclear power stations produce cesium-137. The half-life of cesium-137 is 30 years, which means that it takes over

200 years to reduce it to 1% of its former level. For this reason, an accident at a nuclear power plant can contaminate the environment around for generations, which is why the Chernobyl accident in the Ukraine in 1986 was such an environmental disaster. It released a large amount of radioactive cesium-137 which drifted all over Western Europe, affecting sheep farms as far west as Scotland, Ireland, and Wales, over 1500 miles from the accident. There it was washed to earth by heavy rain and taken up by the roots of plants, thus becoming part of the vegetation that sheep ate.

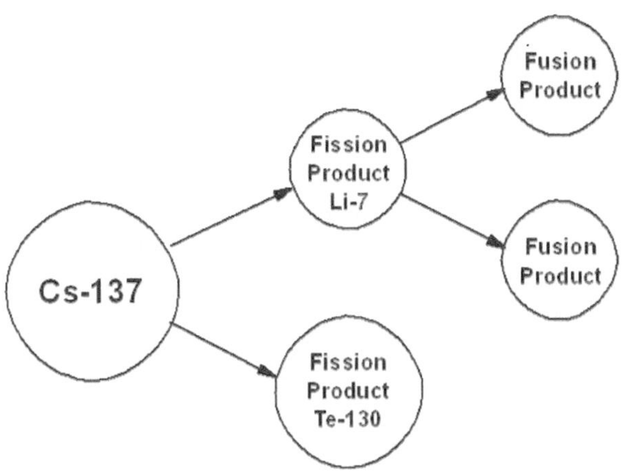

Low energy transmutation of cesium-137

Using low energy transmutation, it may be possible to convert cesium-137 to tellurium-130, a stable non-radioactive isotope, thus redirecting and compressing the cesium-137 decay cycle (see illustration above). The LENR-induced, or low energy fission formula is as follows:

$$^{137}Cs \rightarrow {}^{7}Li + {}^{130}Te$$
Cesium-137 $\rightarrow$ lithium-7 + tellurium-130

In theory, the low energy fission reaction would be triggered by the low energy fusion of lithium and sulfur:

$${}^{7}\text{Li} + {}^{32}\text{S} \rightarrow {}^{39}\text{K}$$
Lithium-7 + sulfur-32 → potassium-39

In a separate experiment, cesium-137 may also transmute into neodymium-148 through a low energy fusion reaction:

$${}^{137}\text{Cs} + {}^{11}\text{B} \rightarrow {}^{148}\text{Nd}$$
Cesium-137 + boron-11 → neodymium-148

If the fusion reaction can be proven and scaled to production levels, it would then be possible to convert dangerous radioactive waste into a valuable rare earth metal widely utilized today in the magnets in hybrid vehicles.

Iodine-129

Iodine-129 is a long-lived isotope of iodine created primarily from the fission of uranium and plutonium in nuclear reactors. It decays with a half-life of 15.7 million years. Significant amounts were released into the atmosphere following nuclear weapons tests in the 1950s and 1960s. Iodine-129 is long-lived and mobile in the environment and is thus of special importance in disposal and management of spent nuclear fuel. It may be possible to compress the natural decay cycle of this radioisotope through the process of low energy fission. The low energy fission reaction is as follows:

$${}^{129}\text{I} \rightarrow {}^{7}\text{Li} + {}^{122}\text{Sn}$$
Iodine-129 → lithium-7 + tin-122

Once again, according to theory, low energy fusion of lithium and sulfur would serve as the catalyst for the reaction:

$${}^{7}\text{Li} + {}^{32}\text{S} \rightarrow {}^{39}\text{K}$$
Lithium-7 + sulfur-32 → potassium-39

During the experiment, iodine-129 may also transmute into barium-146 through a simultaneous fission reaction:

$$^{129}\text{I} + {^7}\text{Li} \rightarrow {^{136}}\text{Ba}$$
Iodine-129 + lithium-7 → barium-136

TECHNETIUM-99

Technetium-99 is radioisotope of technetium that decays with a half-life of 211,000 years to stable ruthenium-99. It is the most significant long-lived fission product of uranium-235. Its high fission yield, relatively long half-life, and mobility in the environment make technetium-99 one of the more problematic components of nuclear waste. There have been releases into the environment from atmospheric nuclear tests, nuclear reactors, and in the late 1990s from the Sellafield plant, which released nearly 1,000 kg into the Irish Sea. It may be possible to accelerate the half–life of Tc-99 by inducing the following low energy fission reaction:

$$^{99}\text{Tc} \rightarrow {^7}\text{Li} + {^{92}}\text{Zr}$$
Technetium-99 → lithium-7 + zirconium-92

Once again, in theory, the reaction would be triggered by the low energy fusion of lithium and sulfur:

$$^7\text{Li} + {^{32}}\text{S} \rightarrow {^{39}}\text{K}$$
Lithium-7 + sulfur-32 → potassium-39

During the experiment, Tc-99 may also transmute into Pd-106 through the following fusion reaction:

$$^{99}\text{Tc} + {^7}\text{Li} \rightarrow {^{106}}\text{Pd}$$
Technetium-99 + lithium-7 → palladium-106

GUIDELINES FOR METHODOLOGY

The experiments on low energy transmutation cited in the Abstract can serve as a starting point for designing experiments to test the nuclear reduction hypothesis presented in this paper. [10] A vacuum tube similar to that used in the QR low energy transmutation tests is shown below and can be considered as standard for the nuclear reduction tests. Because silver is a strong conductor of electricity and a neutron absorber, we propose using it as the anode and cathode material, with other test materials adjusted for each experiment as indicated below. Moreover, silver may react independently with lithium to form tin (^{109}Ag + ^{7}Li $\rightarrow$ ^{116}Sn). This reaction was noted in a previous QR test. [11] Keep in mind that these suggestions are guidelines only, based on previous low energy fusion and fission experiments. They will need to be adjusted in real time based upon further study and experience.

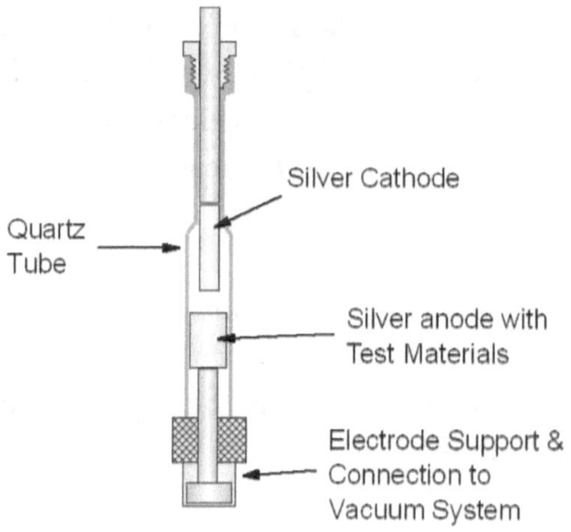

Tube and electrode configuration

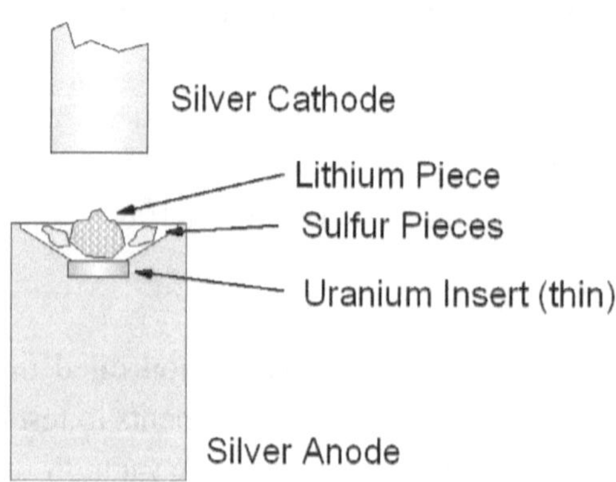

Electrodes and test material suggested for the U-235 $\rightarrow$ Li-7 + Ac-228 experiment

Uraninum-235:

$$^{235}U \rightarrow {}^{7}Li + {}^{228}Ac \rightarrow {}^{208}Pb$$

Electrodes made of Ag
Test Materials:

1. Uranium insert (thin wafer or foil) in anode
2. Lithium test material
3. Sulfur test material
4. Pure neon/oxygen backfill

Plutonium-239:

$$^{239}Pu \rightarrow {}^{11}B + {}^{228}Ac \rightarrow {}^{208}Pb$$

Electrodes made of Ag
Test Materials:

1. Plutonium insert (thin wafer or foil) in anode
2. Boron test material
3. Sulfur test material (optional)
4. Pure neon/oxygen backfill

Radium-226:

$$^{226}Ra \rightarrow {}^{12}C \rightarrow {}^{214}Pb \rightarrow {}^{206}Pb$$

Electrodes made of Ag
Test Materials:

1. Radium insert (thin wafer or foil) in anode
2. Carbon (graphite) test material
3. Sulfur test material
4. Pure nitrogen/oxygen backfill*

*Note: Adding nitrogen allows the process to take advantage of potential carbon-nitrogen reactions such as those noted in QR research. [12]

Cesium-137:

$$^{137}Cs \rightarrow {}^{7}Li + {}^{130}Te$$

Electrodes made of Ag
Test Materials:

1. Cesium insert (thin wafer or foil) in anode
2. Lithium test material
3. Sulfur test material
4. Pure neon/oxygen backfill

$$^{137}Cs + {}^{11}B \rightarrow {}^{148}Nd$$

Electrodes made of Ag
Test Materials:

1. Cesium insert (thin wafer or foil) in anode
2. Boron test material
3. Sulfur test material
4. Pure neon/oxygen backfill

Iodine-129:

$$^{129}I \rightarrow {}^{7}Li + {}^{122}Sn$$
$$^{129}I + {}^{7}Li \rightarrow {}^{136}Ba$$

Electrodes made of Ag
Test Materials:

1. Iodine inserted in or on anode
2. Lithium test material

3. Sulfur test material
4. Pure neon/oxygen backfill

Technetium-99:

$$^{99}\text{Tc} \rightarrow {}^{7}\text{Li} + {}^{92}\text{Zr}$$
$$^{99}\text{Tc} + {}^{7}\text{Li} \rightarrow {}^{106}\text{Pd}$$

Electrodes made of Ag
Test Materials:

1. Technetium insert (thin wafer or foil) in anode
2. Lithium test material
3. Sulfur test material
4. Pure neon/oxygen backfill

Procedure for the Above Experiments:

1. Insert is placed on or into the anode.
2. Measured quantity of test materials are placed in anode recess.
3. Glass/quartz tube is placed over the anode assembly.
4. Cathode is inserted into the tube and secured at the desired separation from the anode.
5. Fill with neon (or nitrogen for Ra-226) to 2 torr.
6. Strike plasma using direct current (DC).
7. Admit oxygen fill to 6 torr. Continue until reaction noticeably slows or tube is in danger of breaking (approximately 10-20 minutes.)
8. Disconnect power and allow sample to cool.

CONCLUSION

As of this writing, the problem of nuclear waste disposal remains unsolved. In an op-ed published in the *Santa Monica Daily Press,* [13] Dr. Jeffrey Patterson, former head of Physicians for Social Responsibility (PSR) stated:

2011 was a scary year for nuclear reactor sites. The summer floods threatened to encroach on reactors in Nebraska and Iowa, an earthquake and a hurricane happened in quick succession to rattle and flood the East Coast, and the continuing events of the Fukushima-Daichi reactor accident provided harrowing examples of the threats posed to spent fuel at reactor sites. The fate of spent fuel there kept the world on edge for days. It's worth noting that the amount of fuel in vulnerable storage pools in Japan was far less than what is crowded into pools at many U.S. reactors. As we all learned, a loss of coolant could produce a fuel melt and large radiation release. It wasn't supposed to be this way. Used reactor fuel was to be permanently stored in deep underground repositories, away from floods and other natural hazards. But the solution to the nation's nuclear waste problem has been elusive for decades. Meanwhile, 65,000 metric tons of spent reactor fuel is still looking for a home.

The Blue Ribbon Commission on America's Nuclear Future proposes transferring spent nuclear fuel, now scattered at 70 locations around the U.S., to temporary storage areas, pending selection of more permanent deep geologic repositories. This proposal is not without controversy. As Dr. Patterson states:

> Moving spent fuel around the country is not a risk worth taking. Rather than addressing the problem, an "interim" facility would only relocate it. So what is the best option? Hardened on-site storage of spent fuel. It's safe, cost-effective—and readily available. PSR and over 170 public interest organizations from all 50 states are calling for adoption of this approach. Storing reactor fuel at reactor sites in hardened buildings that can resist severe attacks, such as a direct hit by high-powered explosives or a large aircraft, as is done in Germany, offers the safest and most sensible option until a permanent repository can be found.

These proposals offer opportunities for research on low energy transmutation. Research laboratories could be set up at future on-site hardened facilities or even now at current waste storage sites, as well as at future interim facilities. These laboratories can begin first-round investigation of low energy transmutation. If successful, scale-up can proceed to levels required to reduce the on-site, regional, and global inventory of nuclear waste. Moreover, low energy transmutation may offer an efficient low-cost alternative to accelerator transmutation of waste (ATW). In 1999, the U.S. Department of Energy's (DOE) Office of Civilian Radioactive Waste Management submitted a report to Congress entitled "A Roadmap for Developing Accelerator Transmutation of Waste (ATW) Technology." Sekazi K. Mtingwa of MIT describes this approach as follows: [14]

Transmutation means the transformation of one atom into another by changing its nuclear structure. In the present context this means bombarding a highly radioactive atom with neutrons, preferably fast neutrons, from either a fast nuclear reactor or spallation neutrons created by bombarding protons from a high-energy accelerator on a suitable target.

The Oak Ridge National Laboratory (ORNL) is currently investigating methods for accelerator transmutation (ATW) of nuclear wastes. An article in the ORNL *Review* states: [15]

Conceived by scientists at Los Alamos National Laboratory, ATW uses a linear accelerator system to produce neutrons for transmutation of excess weapons plutonium and other radioactive DOE wastes, such as technetium-99 and iodine-129. Ultimately, the potential of partitioning and transmutation to waste management is this: If a radioactive waste stream no longer exists, then it poses no radiological hazard. More than anything else, this simple fact has spurred the recent resurgence of interest in partitioning-transmutation technology.

Meanwhile, in the Eurozone, the European nuclear establishment is pressing ahead with a $1.2 billion R&D project to look into high-energy neutron-induced transmutation. The first stage of the project, the setup of a demonstration system known as "Guinevere" that combines a particle accelerator and a nuclear reactor, took place in January 2012 at the Belgian Nuclear Research Center at Mol. A larger version of the reactor system, known as Myrrha (Multipurpose Hybrid Research Reactor for High-tech Applications), is scheduled to become operational in 2023. A press release from the World Nuclear Association explains the thinking behind the project: [16]

> Myrrha will be able to produce radioisotopes and doped silicon, but its research functions would be particularly well suited to investigating transmutation. This is when certain radioactive isotopes with long half-lives are made to "catch" a neutron and thereby change into a different isotope that will decay more quickly to a stable form with no radioactivity. If achievable on an industrial scale, transmutation could greatly simplify the permanent geologic disposal of radioactive waste.

The Quantum Rabbit group estimates that research on low energy transmutation could begin at a fraction of the estimated $1.2 billion startup cost of the Myrrha project. (QR estimates $1.2 million for feasibility study and $12 million to develop a prototype system, amounts that are respectively 0.1% and 1% the cost of Myrrha.) Rather than a highly centralized billion-dollar processing system, low energy transmutation technology could be distributed to nuclear power stations around the globe at an affordable cost. The task of nuclear remediation would become the responsibility of the individual power station and thus remain local instead of becoming highly centralized. Also, the amount of power needed to conduct low energy transmutation would be miniscule compared to the power required to operate a particle accelerator and nuclear reactor. At the very least, research on low energy transmutation should proceed on a parallel track to the high-energy neutron-induced transmutation projects currently underway or under consideration in order to determine which approach yields the most promising results.

NOTES:

[1] Esko, Edward, "Anomalous Metals Part II," *Infinite Energy*, No. 103, 2012.

[2] Esko, Edward and Jack, Alex, *Cool Fusion*, second edition, Amber Waves, Becket, Mass., USA, pp. 56-113, 132-151, 2012.

[3] Esko, Edward, "In Search of the Platinum Group Metals Part II," *Infinite Energy*, No. 104, 2012.

[4] Esko, Edward and Jack, Alex, *Cool Fusion*, second edition, Amber Waves, Becket, Mass., USA, pp. 56-60, 88-97, 2012.

[5] Argonne National Laboratory, "Human Health Fact Sheet," Fig. N.3 Natural Decay Series: Thorium-232, 2005.

[6] Esko, Edward, "In Search of the Platinum Group Metals Part II," *Infinite Energy*, no. 104, 2012.

[7] Esko, Edward and Jack, Alex, *Cool Fusion*, second edition, Amber Waves, Becket, Mass., USA, pp. 56-60, 88-97, 2012.

[8] Yasunari, Teppei J., Stohl, Andreas, Hayano, Ryugo S., Burkhart, John F., Eckhardt, Sabine, Yasunari, Tetsuzo, "Cesium-137 deposition and contamination of Japanese soils due to the Fukushima nuclear accident," Proceedings of the National Academy of Sciences, November, 14, 2011.

[9] Emsley, John, *Nature's Building Blocks: An A-Z Guide to the Elements*, first edition, Oxford University Press, Oxford, England, pp. 82, 2001.

[10] Refer to Occupational Safety and Health Administration (OSHA) guidelines for the handling of hazardous materials prior to initiating these experiments.

[11] Esko, Edward and Jack, Alex, *Cool Fusion*, second edition, Amber Waves, Becket, Mass., USA, pp. 56-113, 132-151, 2012.

[12] Esko, Edward and Jack, Alex, *Cool Fusion*, second edition, Amber Waves, Becket, Mass., USA, pp. 79-87, 2012.

[13] Patterson, Jeffrey, "Time to fix our nuclear waste disposal system," *Santa Monica Daily Press*, January 6, 2012.

[14] Mtingwa, Sekazi K., "Feasibility of Transmutation of Radioactive Isotopes," An International Spent Nuclear Fuel Storage Facility — Exploring a Russian Site as a Prototype: Proceedings of an International Workshop, The National Academies Press, Washington, D.C., 2005.

[15] Michaels, Gordon E., "Partitioning and Transmutation: Making Wastes Nonradioactive," Oak Ridge National Laboratory Review, Vol. 44, No. 2, 2011.

[16] World Nuclear News, World Nuclear Association, January 11, 2012.

Source: Edward Esko, "LENR-Induced Transmutation of Nuclear Waste," *Infinite Energy*, No. 104, 2012.

2

PRELIMINARY RESEARCH ON NUCLEAR REMEDIATION

ABSTRACT

In my paper "LENR-Induced (Low Energy) Transmutation of Nuclear Waste" (*Infinite Energy*, No. 104, 2012), I present possibilities for utilizing low energy nuclear reactions (LENR), also known as low energy transmutations, as a means for converting nuclear materials into non-radioactive elements. The proposed experiments are based on experiments conducted at the Quantum Rabbit (QR) laboratory since 2006. Papers on these experiments have appeared in *Infinite Energy* and have been complied in the book *Cool Fusion*, (Amberwaves, 2011/2012.) In this paper I propose beginning the investigation into possible remediation with experiments on the non-radioactive isotopes of strontium, iodine, and cesium. If these three non-radioactive isotopes can be successfully transmuted using low energy processes, it may be possible to apply this method to transmuting their radioactive isotopes, the nuclear fission products strontium-90, iodine-129, and cesium-137.

TRANSMUTATION OF STRONTIUM

Using low energy transmutation, it may be possible to convert non-radioactive strontium-88 into ruthenium-100. The proposed low energy fusion formula is as follows:

$$^{88}Sr + {}^{12}C \rightarrow {}^{100}Ru$$

Strontium-88 + carbon-12 → ruthenium-100

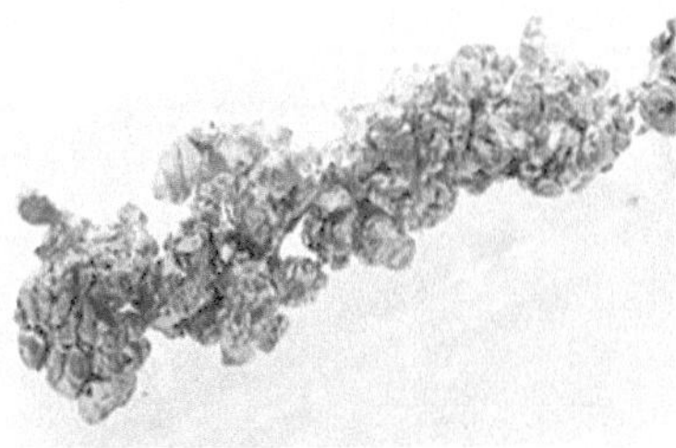

Granules of pure strontium

Using low energy transmutation, it may be possible to convert non-radioactive strontium-88 into ruthenium-100. The proposed low energy fusion formula is as follows:

$$^{88}Sr + {}^{12}C \rightarrow {}^{100}Ru$$

Strontium-88 + carbon-12 → ruthenium-100

The proposed methodology for the test is essentially the same as that deployed in tests conducted by the research team at Quantum Rabbit lab in Owls Head, Maine (see "In Search of the Platinum Group Metals," *Infinite Energy*, No. 92, 2010). The test is conducted in a specially designed vacuum tube with electrodes of pure graphite. Once the graphite electrode (cathode) is secured in the tube a catalyst of pure sulfur is placed in the quartz midsection. A piece of strontium is placed in the center of the reaction zone. Then, a graphite anode is secured in the opposite end of the tube, so that it touches the strontium piece.

The tube is then pumped down to vacuum and backfilled with pure oxygen to about 3.5 torr. Direct current is passed through the electrodes, sufficient to spark plasma and vaporize the test materials. An electric heating tape or coil can be wrapped around the tube for supplemental baseline heat. Samples can be analyzed by ICP to test for the presence of ruthenium isotopes.

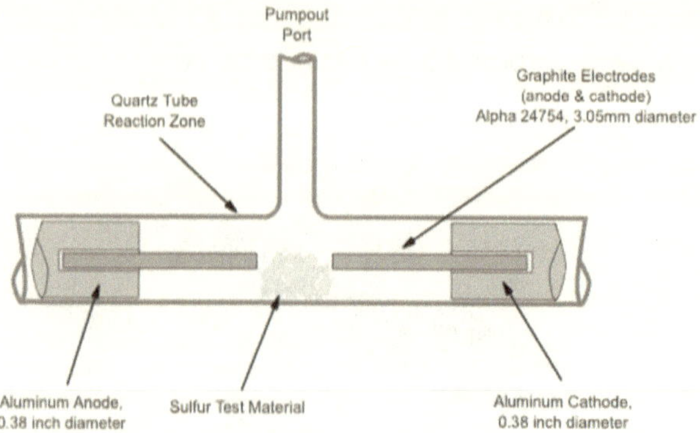

In the proposed strontium test, strontium granules are mixed with the sulfur test material

TRANSMUTATION OF IODINE

Pure iodine crystals

The goal of the iodine study is to transmute the stable isotope of iodine (I-127) into tin (Sn-120) and barium (Ba-134) through low energy nuclear reactions as another step in demonstrating the possibility of the remediation of nuclear materials through low energy transmutation.

The first reaction is a low energy fission reaction:

$$^{127}I \rightarrow {}^{7}Li + {}^{120}Sn$$

Iodine-127 $\rightarrow$ lithium-7 and tin-120

The fission reaction is potentially triggered by the low energy fusion reactions:

$$^7\text{Li} + {}^{32}\text{S} \rightarrow {}^{39}\text{K}$$
Lithium-7 + sulfur-32 → potassium-39

$$^{127}\text{I} + {}^7\text{Li} \rightarrow {}^{134}\text{Ba}$$
Iodine-127 + lithium-7 → barium-134

This test can be conducted in the vertical tube described in the experiments. In the test, a hole is drilled in the center of the lower copper electrode and a pellet of pure lithium is pressed into it. Iodine and sulfur test material are placed on top of the lower electrode. The glass/quartz tube is placed over the electrode assembly. The upper copper electrode is inserted into the tube and secured at the desired separation from the lower electrode and test materials. Oxygen or neon is backfilled to approximately 2 torr and plasma is struck using direct current.

The reaction zone can be heated with a hand torch or an electric heating tape or coil can be wrapped around the tube for supplemental baseline heat, with the goal being to achieve the relatively low melting temperatures of the elements utilized or sought after in the experiment: iodine, lithium, barium, and tin. Their melting temperatures are as follows:

Element	Melting Temperature °C
Iodine	114
Lithium	181
Tin	232
Barium	727

The experiment can continue until the reaction noticeably slows or the tube is in danger of breaking. Samples can be analyzed by ICP to test for the possible transmutation products tin, barium, and potassium. Further refinement of the testing process could demonstrate variations in isotopic composition of the transmutation products. Barium-134, the hypothetical transmutation product of iodine-127 and lithium-7 makes up only 2.4% of naturally occurring barium isotopes. If test samples were found to contain barium-134 in amounts significantly greater than this, it would help rule out contamination as the source of the barium and offer convincing proof of transmutation.

Transmutation of Cesium

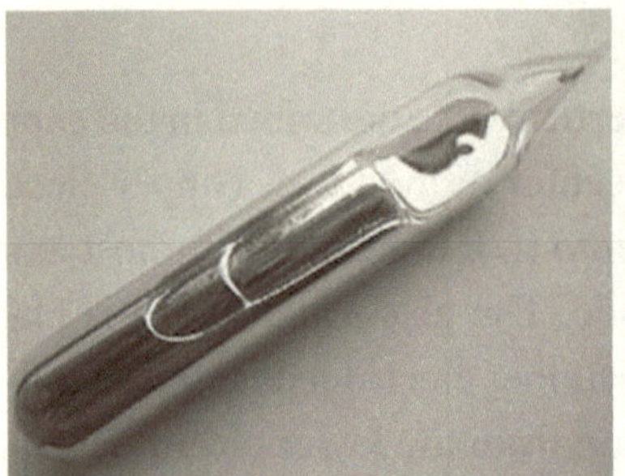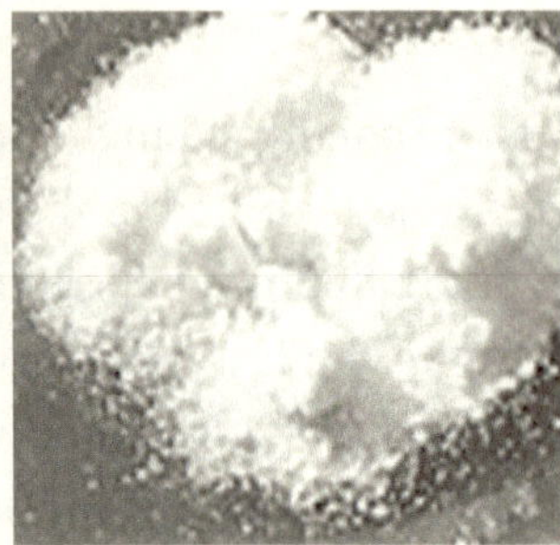

Cesium metal (left) and cesium chloride (right)

Cesium metal is highly reactive. In addition to igniting spontaneously in air, it reacts explosively with water even at low temperatures, more so than other members of the first group of the periodic table that includes lithium, sodium, and potassium. The reaction with solid water occurs at temperatures as low as −116 °C (−177 °F). Because of its high reactivity, the metal is classified as a hazardous material. It is stored and shipped in dry saturated hydrocarbons such as mineral oil. Similarly, it must be handled under inert gas such as argon. It can be stored in vacuum-sealed borosilicate glass ampoules. (In quantities of more than about 100 grams, cesium is shipped in hermetically sealed, stainless steel containers.) In addition, cesium has a melting point of 28.4 °C (83.1 °F), making it one of the few pure metals which are liquid near room temperature. (Mercury is the only metal with a melting point lower than cesium.)

Due to these concerns, the team at Quantum Rabbit is not equipped to handle cesium in its elemental form. Accordingly, it recommended that research begin with one of the less volatile cesium compounds such as cesium chloride (CsCl.) Cesium chloride powder has a melting temperature of 646 °C, thus making it easier to handle than elemental cesium. The goal of the cesium study is to transmute the stable isotope of cesium (Cs-133) into tellurium (Te-126), cerium (Ce-140), and neodymium (Nd-143 and

Nd-144) through low energy fission and low energy fusion reactions. The low energy fission reaction we propose testing is as follows:

$^{133}Cs \rightarrow {}^{7}Li + {}^{126}Te$
Cesium-133 $\rightarrow$ lithium-7 and tellurium-126

Our hope is to trigger the fission reaction by low energy fusion reactions:

7Li + 32S $\rightarrow$ 39K
Lithium-7 + sulfur-32 $\rightarrow$ potassium-39

133Cs + 7Li $\rightarrow$ 140Ce
Cesium-133 + lithium-7 $\rightarrow$ cerium-140

In a separate experiment, we propose testing the possibility of low energy fusion between cesium and boron:

$^{133}Cs + {}^{11}B \rightarrow {}^{144}Nd$
Cesium-133 + boron-11 $\rightarrow$ neodymium-144

These tests can be conducted in the vertical tube described in the experiments. In the first test, a hole is drilled in the center of the lower copper electrode and a pellet of lithium metal is pressed into it. Sulfur crystals and cesium chloride power are sprinkled on the electrode surface. In the second test, cesium chloride powder and boron crystals are placed on the lower electrode surface. Following placement of test materials, the glass/quartz tube is placed over the electrode assembly. The upper copper electrode is inserted into the tube and secured at the desired separation from the lower electrode and test materials. Oxygen or neon is backfilled to approximately 2 torr and plasma is struck using direct current.

As in the previous tests, the reaction zone can be heated with a hand torch or an electric heating tape or coil can be wrapped around the tube for supplemental baseline heat. The experiment can continue until the reaction noticeably slows or the tube is in danger of breaking. Samples can be analyzed by ICP to test for the possible transmutation products tellurium, neodymium, cerium, and potassium.

Source: Edward Esko, "Preliminary Research on Nuclear Remediation, *Infinite Energy*, No. 110, 2013.

3

Appearance of Barium in Lithium-Iodine Plasma

Abstract

In studies conducted at Quantum Rabbit (QR) lab in Owls Head, Maine on April 10, 2013, funded by Woodland Energy and the New Energy Foundation, independent analysis of test samples by inductively coupled plasma mass spectroscopy (ICP-MS) revealed the anomalous presence of barium (Ba). Isotope distribution studies further revealed the presence of barium isotopes at variance with their natural distribution, notably elevated values for barium-134. The vacuum discharge tests employed copper (Cu) electrodes and lithium (Li) and iodine (I) test material. The appearance of novel isotope distribution in test samples diminishes the possibility that these results are due to contamination. Test results raise the possibility that barium was newly created through low energy transmutation.

Background

In my paper "Preliminary Research on Nuclear Remediation" (*Infinite Energy,* No. 110, 2013), I proposed beginning research into possible nuclear remediation with low energy tests on non-radioactive isotopes of strontium, iodine, and cesium, with the concept being that if these non-radioactive isotopes are successfully transmuted, it may be possible to apply the same or similar low energy processes to the transmutation of their radioactive cousins. If scalable, low energy transmutation could offer a sustainable solution to the nuclear waste problem. As I stated in my paper: "The goal of the iodine study is to transmute the stable isotope of iodine (I-127) into barium (Ba-134) through low energy nuclear reactions as another step in demonstrating the possibility of the remediation of nuclear materials through LENR." I proposed conducting tests on the following low energy fusion formula:

$$^{127}I + {}^{7}Li \rightarrow {}^{134}Ba$$

Iodine-127 + lithium-7 → barium-134

I went on to state: "Barium-134, the hypothetical transmutation product of iodine-127 and lithium-7 makes up 2.4% of naturally occurring barium isotopes. If test samples were found to contain barium-134 in amounts greater than this, it would help rule out contamination as the source of barium and offer proof of transmutation." In Test 2 conducted on April 10, barium-134 was found at 4.04%, nearly double the quantity found in naturally occurring barium.

THE LITHIUM-IODINE STUDY

The procedure of the study was exceedingly simple. We conducted three experiments using the vertical vacuum tube deployed in previous tests and shown below. Copper electrodes were inserted in the upper and lower ends of the tubes. A lithium plug was inserted in the center of the lower electrodes in Tests 1 and 2. Pure iodine crystals were placed on top the lithium inserts. In Test 3, for variation, no lithium plug was used. Instead a small piece of lithium was placed in the center of the lower copper electrode. Iodine crystals surrounded the lithium. The three electrode configurations are also shown below.

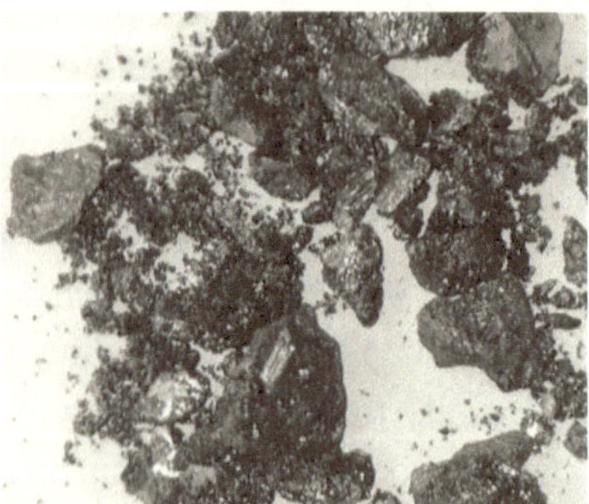

Lithium rod (left) and iodine crystals (right)

After pumping down to 3.5 torr, oxygen was admitted and the power turned on. When the arc was established and plasma struck, additional heat was provided by a hand-held torch. The arc was maintained for approximately 10 to 15 minutes, at which time the power was disconnected and the tubes allowed to cool. Below is the worksheet I prepared prior to the experiments.

Lithium-Iodine Study
April 10, 2013

Inputs:
Copper electrodes
Lithium test material
Iodine test material
Oxygen fill

Procedures:

1. Insert Li plug in lower Cu electrode.
2. Place I on surface of Li plug (do not cover completely.)
3. Position electrode in tube.
4. Maneuver upper electrode into position.
5. Pump down to approx. 3.5 torr.
6. Admit O_2 fill and strike plasma
7. Apply torch as needed (optional.)
8. Continue approx. 15 minutes.

Note: In test 3, a piece of Li is placed on the lower electrode together with I, rather than inserting the Li plug in the electrode.

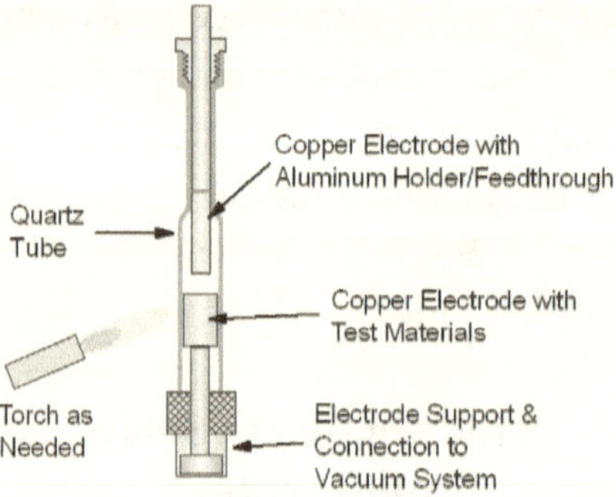

The vertical tube used in the April 10 experiments

All three tests proceeded according to plan. We adjusted the electrode polarity between Test 1 and Tests 2 and 3. In Test 1, the upper electrode served as the anode, and in Tests 2 and 3, the upper electrode served as the cathode. The power supply was the same as that described in the earlier papers.

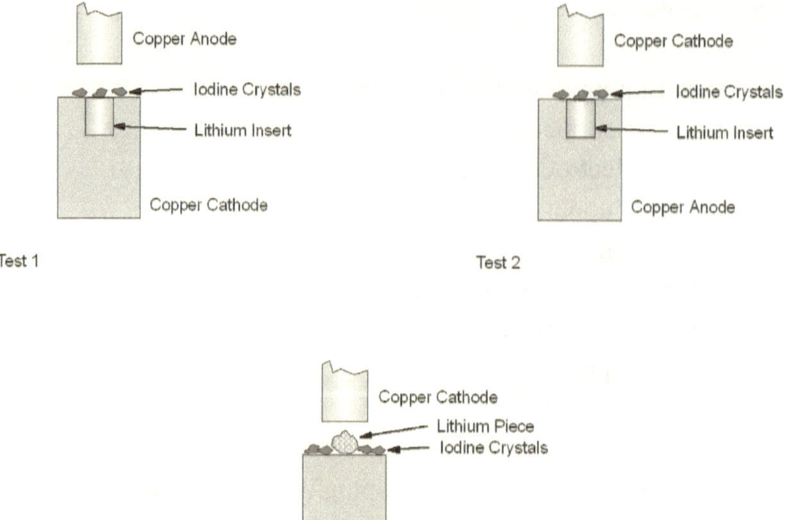

Tests 1 and 2 (above) used a lithium insert. Test 3 used a
piece of lithium surrounded by iodine crystals.

Three sets of samples were collected, labeled, and packed for shipping. The samples were sent to Northern Analytical Laboratory in Londonderry, NH with the following instructions:

Guidelines for Sample Analysis
April 2013

Sample Materials:

A. Test 1: One thick copper electrode (with lithium insert and lithium-iodine residue): one thin copper electrode; and one glass tube.
B. Test 2: One thick copper electrode (with lithium insert and lithium-iodine residue): one thin copper electrode; and one glass tube.
C. Test 3: One thick copper electrode with lithium-iodine residue; one thin copper electrode; and one glass tube.

Requested Analysis by ICPMS:
Barium (isotope distribution percentage)
Cesium

Procedure for Each Sample:

1. Scrape surface (top and sides) of Cu electrodes, including surface of Li insert.
2. Collect powder and flake residue (loose material) from plastic baggie.
3. Scrape residue from inner surface of tube.
4. Combine scrapings and loose material into one sample for analysis.

Note: Each sample to be analyzed separately.

ISOTOPES OF BARIUM

Results came back on May 6. As predicted, barium was found in all three of the test samples, albeit in microscopic quantities. In Test 1 barium was found at 3.5 ppm, in Test 2 at 1.8 ppm, and in Test 3 at 463 ppm (Table

1). According to the Certificates of Analysis from Alfa Aesar, the supplier of the pure elements used in the test, barium was listed at <0.0005 ppm in the copper used for the electrodes. The lithium used in the test showed no value for barium.

Table 1. Values of Barium Before and After Experiment

Starting Concentration (ppm)*	Final Concentration (ppm)**
0.43	3.5 (Test 1)
0.43	1.8 (Test 2)
0.43	463 (Test 3)

*Source: Test Report from Northern Analytical Laboratory, 6/19/13
**Source: Test Report from Northern Analytical Laboratory, 5/6/13

Table 2. Iodine Certificate of Analysis

Product No.:	00158
Product:	Iodine, crystalline, 99.99+% (metals basis)
Lot No.:	D22X020
Purity	99.99%
Nonvolatile matter	0.004%
Chlorine and bromine (as Cl)	<0.005%

To provide a further control, we sent a sample of the iodine used in the tests to Northern Analytical Laboratory with instructions to test for barium. The control sample was taken from the same batch used in the experiments. The test results, shown in Table 2, came back showing traces of barium in the iodine at 0.43 ppm, far below the amounts detected in the test samples. For example, in Test 1, the amount of barium was 8 times greater than that detected in the control, in Test 2, more than 4 times greater than the control, and in Test 3, barium was detected at a level more than 1,000 times greater than that detected in the control (Table 1).

The distribution of barium isotopes in the test samples varied somewhat from the distribution found in nature, suggesting the possibility that detected barium was newly created through low energy transmutation and not introduced through contamination. A comparison of the isotope distribution of natural barium and the barium found in the April experiments is shown in Table 3.

Table 3. Comparison of Barium Isotopes in Nature* and in the April 10 Experiments**

Isotope	Naturally Occurring %	Test 1 %	Test 2 %	Test 3 %
Ba132	0.101	1.36	3.92	0.19
Ba134	2.417	2.73	4.04	2.42
Ba135	6.592	6.62	6.18	6.72
Ba136	7.854	8.17	8.68	8.04
Ba137	11.232	10.9	10.9	11.9
Ba138	71.698	70.2	66.3	70.8

*Source: National Nuclear Data Center
**Source: Test Report from Northern Analytical Laboratory

The greatest variance in isotope distribution was seen in Test 2, with Ba-132 appearing at 3.92% compared to 0.101% in nature; Ba-134 at 4.04% compared to 2.417%; and Ba-138 at 66.3% compared to 71.698%.

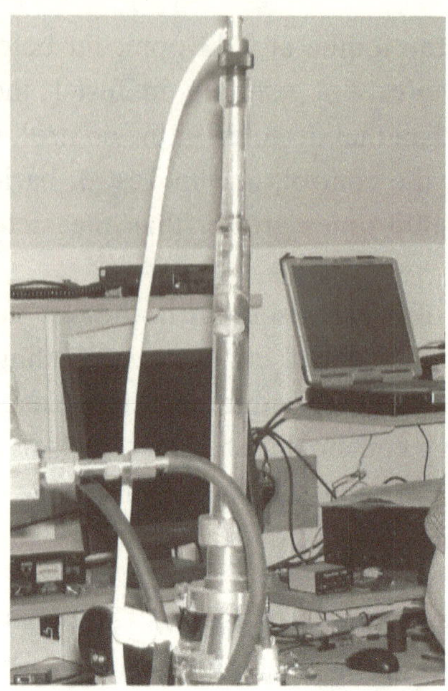

Vertical tube used in the lithium-iodine study

INTERPRETATION

Low energy fusion between iodine (atomic number 53) and lithium (atomic number 3) could explain the consistent appearance of barium (atomic number 56) in all three tests, with a peak at 463 ppm in Test 3. The distribution of barium isotopes is perhaps more difficult to understand. Ba-134 can be accounted for by the fusion of iodine-127 and lithium-7. That formula was predicted beforehand. Barium-132 may have arisen following the ejection of a neutron at the moment of fusion between Li-6 and I-127. Perhaps the low energy fusion process releases neutrons, some of which are captured by newly formed barium-134 nuclei, thus explaining the formation of isotopes heavier than Ba-134. More study is needed to fully explain the origin of the anomalous isotopes found in these experiments.

Source: Edward Esko, "Appearance of Barium in Lithium-Iodine Plasma," *Infinite Energy*, No. 111, 2013.

4

THE BAKKEN: RADIOACTIVE DUMPSITE?

An April 2014 article in the *Wall Street Journal* entitled "Radioactive Waste Is North Dakota's New Shale Problem" describes the growing problem of nuclear waste in the Bakken shale oil formation, the epicenter of the U.S.'s surge in energy production:

> At a deserted gas station in a remote North Dakotan town, local officials recently found an unintended byproduct of the shale-oil boom: hundreds of garbage bags filled with mildly radioactive waste. These bags, which were discovered late February in Noonan, N.D., contained what are known as "oil socks": three-foot-long, snake-like filters made of absorbent fiber. The shale-oil industry uses the socks to capture silt from wastewater resulting from hydraulic fracturing. Days earlier, a similar trove had been found on flatbed trailers near a landfill in Watford City—which, like Noonan, is located in the state's sparsely populated westernmost reaches where the Bakken oil shale formation lies. The two recent incidents show that North Dakota's regulators have been slow to address repercussions from the surge in crude output, ranging from widespread flaring of natural gas at oil wells to drill rigs popping up on historic lands.

Most of the radioactive material in oil socks comes from silt filtered in the process of pumping wastewater down injection wells. Radium, found in soil, rock and water, accumulates in the filtered silt.

"Before the Bakken oil boom we didn't have any of these materials being generated," said Scott Radig, the state's director of waste management. "So it wasn't really an issue."

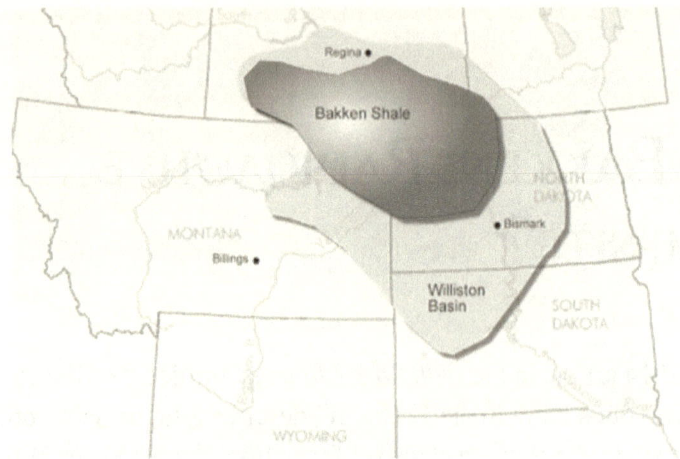

The Bakken shale oil deposit

Confidential sources have told me that the levels of radioactive waste in the Bakken are actually far higher than those reported in the media. If accurate, the U.S. government and the energy interests profiting form the Bakken formation need to address this issue. Research on low energy transmutation, such as that conducted by the author, together with Alex Jack and Woody Johnson, members of the Quantum Rabbit research group, may offer a solution to this growing environmental problem. Radium-226 seems to be the key radioactive contaminant in the Bakken fields. Radium-226 is part of the urianium-238 (U-238) decay chain with a half-life of 1,600 years. With low energy transmutation, it may be possible to compress this time frame considerably by achieving the low energy fission of Ra-226.

Quantum Rabbit research on carbon-arc may offer a method for achieving this possibility. Numerous low energy transmutations have been reported, both in open air and under vacuum. Low energy fusion reactions could possibly be used to prompt the low energy fission of radium-226, compressing the half-life of radium and accelerating the natural decay cycle from more than 1,600 years to approximately 22 years.

The low energy fission reaction we propose testing is as follows:

$$^{226}\text{Ra} \rightarrow {}^{12}\text{C} + {}^{214}\text{Pb}$$
Radium-226 → carbon-12 + lead-214

This low energy fission reaction could possibly be triggered by low energy fusion reactions such as those between carbon and oxygen noted in QR carbon-arc research:

$$^{12}\text{C} + {}^{16}\text{O} \rightarrow {}^{28}\text{Si}$$
Carbon-12 + oxygen-16 → silicon-28

$$_2(^{12}\text{C} + {}^{16}\text{O}) \rightarrow {}^{56}\text{Fe} \ ^{+2 \text{ protons}}$$
$_2$(Carbon-12 + oxygen-16) → iron-56 + two protons

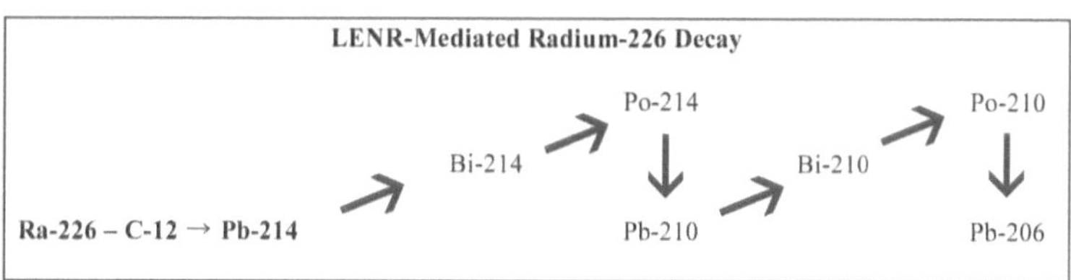

Downward arrows represent alpha decay; upward arrows beta decay.

The Quantum Rabbit research group, in association with Planetary Health and associates, hopes to begin research on the remediation of the nuclear waste in the Bakken and at other sites around the world.

Source: Edward Esko, "The Bakken: Radioactive Dumpsite?" *Dandelion Essays*, 2016.

When one is confronted with a fact which is in opposition
with a prevalent theory, one must accept this fact and
abandon the theory, even though the latter, supported
by great men, may be generally subscribed to.
CLAUDE BERNARD

Part IV

In Search of Nanonovae

Platinum is much rarer than both gold and silver—so rare,
in fact, that all of the platinum ever mined could fit into
your living room. And palladium is even rarer than that.
OUTSIDER CLUB

It is premature to reduce the vital process to the
quite insufficiently developed conceptions of 19th
and even 20th century physics and chemistry.
L. DE BROGLIE

INTRODUCTION

In the microscopic *nanoverse*, as well as in the universe at large, no two things are identical. Common sense confirms this to be true. However, this is in contradiction to the view of modern science. In today's science, all protons are identical to one another, all electrons are identical to one another, and all neutrons are identical, as are any and all subatomic particles. It follows then that all the atoms in a given element must also be identical. All the atoms of oxygen in the atmosphere must be the same as all the others in the atmosphere; every gold atom in a ring must be identical to every other atom of gold in the ring; and any one carbon atom in a grain of brown rice will be identical to all the other atoms of carbon in that grain.

However, if we accept that nothing is identical, in reality, no two atoms or subatomic particles can ever be exactly the same. No two electrons are exactly alike. No two neutrons are carbon copies of one another. No two protons are identical. Atoms and the subatomic particles that make them up are like snowflakes, leaves, human voices, or fingerprints. They share an underlying pattern, yet each is completely unique. So the atoms and subatomic particles in Mary's gold ring are not identical to the atoms and subatomic particles in Jennifer's gold ring. Nor are any two atoms or subatomic particles in each of the rings exactly the same. The principle of nonidentity applies to all phenomena, from the microcosmic level of atoms and preatomic particles, to the macrocosmic level of galaxies, super clusters of galaxies, and so on throughout the universe.

The protons that occupy the nucleus of the atom carry a positive electric charge. The electrons orbiting the periphery are negatively charged. Neutrons carry no charge and are electrically neutral. As we know, two positively charged objects repel each other. Two negative charges will also repel. A positive charge attracts a negative charge, and vice versa. According to this law, under conditions such as those found on earth, atomic nuclei, which derive their charges from positively charged protons, will always repel each other. Nuclei are thus prevented from fusing with other nuclei.

As modern physics points out, the force of electric repulsion existing between nuclei is known as the Coulomb barrier. According to standard physics, this barrier can only be overcome under the most extreme conditions, such as extremely high temperatures, pressures, and energies. It requires energy on an order of magnitude equal to a nuclear explosion to overcome the barrier and force atoms to fuse. That is why, according to the standard model, low energy transmutation is impossible. Keep in mind however, that that assumption is predicated on the notion that all protons and other subatomic particles are identical.

However, the principle of non-identity reveals that the Coulomb barrier is relative, not absolute. If no two protons are identical, some degree of attraction, however slight, must exist between them. It may be possible to enhance that attraction, perhaps by manipulating external conditions such as heat, pressure, and electrical energy, in a manner sufficient to weaken the barrier and allow a certain percentage of nuclei to fuse. Also throughout the universe, what has a front has a back. As the front becomes larger and more powerful, so does the back. This principle was also expressed by Isaac Newton: "Every action has an equal and opposite reaction." So, when two nuclei are repelled by one another, they are also attracted. The more strongly they are repelled, the stronger the potential for attraction.

As powerful as the repulsive force of the Coulomb barrier appears to be, existing right behind that repulsion is an equally powerful attractive force. Once again, by manipulating external conditions, it may be possible to "flip" the repulsive Coulomb force into an equally powerful attractive force, thus allowing nuclear fusion to occur. Together with the attraction of opposites and repulsion of likes, there is another process we must consider. According to the theorem: a large positive charge attracts a smaller positive charge, while a large negative charge attracts a smaller negative charge. This may help us understand how two atomic nuclei, both of which contain positively charged protons, could fuse into larger nuclei in a process of transmutation. Let us take an example from an actual experiment. As we see a nucleus of lithium, shown on the right, has 3 protons and 4 neutrons.

Sulfur, at left, has 16 protons and 16 neutrons. The number of protons and neutrons determines the atomic weight of an element.

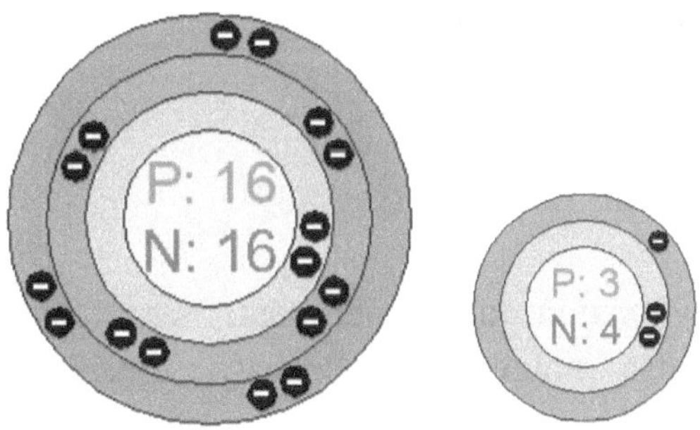

Sulfur (left) and lithium (right)

The number of electrons orbiting the nucleus determines the atomic number. The number of positively charged protons is usually equal to the number of negatively charged electrons. Thus the atom is electrically neutral. If we add the 3 protons and 4 neutrons in the lithium nucleus, we get an atomic weight of 7. If we add the 16 protons and 16 neutrons in the sulfur nucleus, the atomic weight comes out to 32. That means that a nucleus of sulfur is more than 4 times heavier than a nucleus of lithium, or that sulfur has a positive charge that is 4 times greater than that of lithium. Sulfur thus qualifies as a "large" positive, lithium as a "smaller" positive. If the theorem that big attracts small is correct, there will be attraction between lithium (small positive) and sulfur (large posi-tive.) That attraction may offset the Coulomb barrier to some degree and explain the apparent fusion of lithium and sulfur witnessed in several experiments.

In ten experiments conducted between 2009 and 2012, the Quantum Rabbit (QR) research team placed lithium and sulfur on a copper electrode under vacuum. The electrode was charged with enough electricity to generate plasma. When the experiment was finished, test material was collected and sent to an outside lab for analysis. In each of the ten experiments, analysis revealed the presence of potassium (K), although, aside from minute traces in the quartz tube, no potassium was introduced into the test.

The values for potassium ranged from a low of 3 parts per million to a high of 750 parts per million (refer to *Corking the Nuclear Genie*, by Edward Esko and Alex Jack, Amberwaves Press 2014.) The QR researchers were testing the following formula:

$$^{7}Li + {}^{32}S \rightarrow {}^{39}K$$
Lithium-7 + sulfur-32 $\rightarrow$ potassium-39

In our low energy vacuum tube experiments, we attempt to create conditions similar to those in outer space. Polarity animates the flow of energy in the tube as well as in the galaxy. Electric current passes between anode and cathode, separated by several inches. Similarly, electric current passes between the oppositely charged poles of the galaxy, separated by 100,000 light years. It may be that in this highly charged plasma sphere, elements are continually forming in a never-ending process of transmutation. It may also be that in the QR tube, also containing highly charged plasma, transmutation is taking place on an infinitely smaller scale. Microcosm equals macrocosm. As is above so is below. Skeptics often cite contamination as the source of the purported transmutation products reported in experiments. When Alex Jack, Woody Johnson and I visited Dr. Peter Haglestein at MIT several years ago, Dr. Haglestein, who has several decades of experience with LENR research, warned that critics would always cite contamination, no matter how convincing the evidence.

As Einstein famously said, "If the facts don't fit the theory, change the facts." So in regard to low energy transmutation, the appearance of anomalous elements in the test results doesn't conform to the standard model of physics; therefore, the results can only be due to contamination, not transmutation. In many of our studies, the amount of purported transmutation products is extremely small, thus results could be attributed to contamination. However, in some of our studies, the amounts of purported transmutation products, including rare metals, were simply too large to be attributed to contamination. Moreover, these products were accurately predicted beforehand.

We may be on the cusp of a huge change in the way we view the world. A new model of element formation could be a critical step in that process.

A new era may now be beginning. It may be that elements are forming through a process that is not clearly understood. The process of creation may be occurring throughout the universe at this very moment. Creation may simply be part of the peaceful and orderly process of change that is without beginning or end, and which inhabits a realm beyond time and space.

Edward Esko
Pittsfield and Lenox, MA

A continual process links the elements. They
are part of an evolutionary continuum.
MICHIO KUSHI

The abundance of palladium in the earth's crust is estimated
to be about 1 to 10 parts per trillion. That makes it one
of the ten rarest elements found in the earth's crust.
ENCYCLOPEDIA.COM

1

In Search of Nanonovae

In his article entitled, "Was Transmutation Observed at the Quantum Rabbit Laboratory?" (*Infinite Energy* issue 92), physicist Matthias Grabiak asked the question: "How likely is it that a small laboratory in New Hampshire with simplistic experimental set-ups was able to achieve nuclear reactions that are thought to be only possible under such extreme conditions as found in supernova explosions?" Grabiak was questioning the low energy transmutations reported to have taken place in the Quantum Rabbit laboratories from 2007 to 2013.

Grabiak is using as reference the popular theory that heavy elements such as palladium, platinum, and gold are formed in the explosions of dying stars known as supernova, or in the collision of "neutron" stars left over from supernova explosions. The latest such collision, thought to have been witnessed in 2017, produced a new star known as a "kilonova." The star faded after several days, but, according to the hypothesis, its high temperatures and pressures caused heavy elements, such as gold, to form. Astronomers believed they witnessed the formation of gold on a cosmic scale, claiming that "10-earth's" worth of gold was ejected into space.

The modern theory of element formation is based on the belief in a cosmic "chain of violence" that started with the big bang (mostly hydrogen and helium), and continued through to element creation in stars (up to iron), and peaking in supernovae, the sudden violent explosion of a star as the result of high temperature—6,000 times hotter than the sun—and pressures (iron through to uranium.) Not only are the events in this hypothesis centered on violence, they are somewhat fixed and static, and centered on the "death" of stars, and not on the endless process of creation that is life itself. According to Britain's Astronomer Royal Sir Martin Rees, "We are literally the ashes of long dead stars."

So as to provide an alternative to the hypothesis that the elements were created in a chain of violence and death, the research team at Quantum Rabbit LLC has been conducting experiments on low energy transmutation. These experiments are attempting to duplicate on the microscopic scale events that may be occurring on the macrocosmic scale, specifically, the ongoing process of element formation. These experiments suggest it may be possible to create what we refer to as *nanonovae* (as opposed to *supernovae*): microscopic events requiring relatively little in the way of temperature, pressure, and energy and out of which elements are formed.

In an experiment conducted in 2016, pure tungsten powder was placed in a graphite crucible together with small pieces of pure boron. The crucible was attached to a clip that connected it to a power pack consisting of four 12-volt batteries. A carbon rod was attached to a clip connecting it to the opposite pole of the battery pack. The rod was brought into proximity with the tungsten and boron sitting at the bottom of the open crucible. An electric spark, similar to a lightning strike, resulted. The carbon arcing process was repeated about fifty times.

Upon ICP analysis at Northern Analytical Laboratory, the residue at the bottom of the crucible was found to contain gold in the amount of 3,000 ppb (parts per billion) or 3 ppm (parts per million.) The relative abundance of gold in the earth's crust is estimated to be 0.004 ppm, substantially less than that found in the experiment. The experiment yielded a significant signal (3 ppm) to noise (0.004 ppm) ratio. The formula being tested in the experiment was as follows:

$$^{11}B + {}^{186}W \rightarrow {}^{197}Au$$
(Boron-11 + tungsten-186 → gold-197)

In another experiment, conducted at M & M Glassblowing in Nashua, NH, in 2018, the research team placed pure zinc powder in a vertical vacuum tube and mixed it with pieces of pure sulfur. The materials were placed on the surface of a copper electrode that was used as the lower anode. A copper cathode was positioned above the anode and the electrodes and materials were sealed in a quartz tube. The tube was pumped down to 3.5 torr and about 50 amps of direct current were used to electrify the anode and cathode. A glow discharge was produced, vaporizing both the zinc powder and sulfur pieces. After about 15 minutes the electricity was shut off and the tubes allowed to cool.

ICP analysis at Northern Analytical Laboratory found palladium at 2,000 ppb (2 ppm.) Palladium is an incredibly rare element. According to *Encyclopedia.com*, palladium is found in the earth's crust at only 1 to 10 parts per trillion (ppt.) If that estimate is accurate, the experiment resulted in another significant signal to noise ratio: 2,000,000 ppt versus 1 to 10 ppt. The experiment was designed to test the following formula:

$$^{34}S + {}^{68}Zn \rightarrow {}^{102}Pd$$
(Sulfur-34 + zinc-68 → palladium-102)

If elements are indeed being created in nanonovae, what process can help explain this phenomenon? The explanation may be as simple as observing what happens when we throw two stones into a pond, one after the other.

The "ripple-wave" model of low energy transmutation

The two separate ripples amplify and merge into a larger ripple. In the *nanonova* matrix, under the prompting of directed heat and energy, nuclei may undergo a similar transformation. Two lighter nuclei amplify and expand, dissolving their identities and merging into a new larger and heavier nucleus.

Source: Edward Esko, *In Search of Nanonovae*, IMI Press, 2019

2

QUANTUM RABBIT PRECEDENTS

We trace our lineage in the field of low energy transmutation (see *Cool Fusion* and *Corking the Nuclear Genie*, Amberwaves Press 2012 and 2014) to the work conducted by George Ohsawa, Louis Kervran, and Michio Kushi in the 1950s and 1960s, in Tokyo, France, and Cambridge, Mass. However, readers might be interested to know that modern research on low energy transmutation actually took place four decades earlier at Tokyo University in Japan. Coincidentally, Tokyo University is the Alma matter of Michio Kushi, who, in the 1970s, introduced the Quantum Rabbit partners to low energy transmutation through his Boston lectures. In 1924 and 1925, Professor Hantaro Nagaoka and his associates at Tokyo University conducted some 200 experiments.

Professor Hantaro Nagaoka

Nagaoka used high-current electric arc discharges between tungsten electrodes immersed in liquid hydrocarbon transformer oil in which they detected the successful transmutation of tungsten into visible flecks of gold and platinum. In June 1925, Nagaoka went a world tour in which he spoke to scientific and lay audiences about his transmutation experiments and handed out samples with tiny bits of gold that had been created. A letter on his work was published in the journal *Nature* in July 1925, in which he encouraged other scientists to try to duplicate his results. Unfortunately, modern nuclear physics preempted Nagaoka's work and his results were not followed up on, until the independent work of Ohsawa, Kervran, and Kushi four decades later. The Quantum Rabbit research of the 2000s was based on this independent work.

If valid it seems Prof. Nagaoka achieved transmutation of tungsten into gold through a process of low energy *fusion*. Before hearing of Nagaoka's work, I developed a formula for this possible low energy fusion reaction: $^{11}B + ^{186}W \rightarrow ^{197}Au$ (boron-11 + tungsten-186 $\rightarrow$ gold-197.) In experiments conducted in 2008, 2009, and 2011, the QR team apparently produced nano-quantities of gold through a process of low energy *fission*. This formula may help explain Nagaoka's results. Electric arcing may have caused atoms of carbon ^{12}C to shed a proton and neutron and transmute into boron ^{11}B. The boron may have then fused with tungsten to form gold, as described in the above formula.

Our experiments involved the low energy fission of lead into gold through the following formula: $^{204}Pb \rightarrow ^{7}Li + ^{197}Au$ (lead-204 $\rightarrow$ lithium-7 + gold-197.) In our theory, the low energy fission of lead into lithium and gold is triggered by a low energy fusion reaction: $^{7}Li + ^{32}S \rightarrow ^{39}K$ (lithium-7 + sulfur-32 $\rightarrow$ potassium-39.) In all experiments, potassium was detected in the end product. However, unlike the results reported by Nagaoka, the microscopic quantities of gold in the QR tests were not visible to the naked eye, but were detectable only through inductively coupled plasma spectroscopy (ICP.)

We are currently working on proposals for a new round of experiments on low energy transmutation, including studies of foundation formulas, rare earth metals, rare metals, and remediation of nuclear materials. We welcome inquiries and are seeking venture partners. We would like to acknowledge Prof. Nagaoka's pioneering work and hope to confirm and develop it.

Source: Edward Esko, "Quantum Rabbit Precedents," *Dandelion Essays,* 2016

3

LOW ENERGY TRANSMUTATIONS AND THE FORMATION OF ELEMENTS

The following is from email correspondence between Matthias Grabiak (Was Transmutation Observed at the Quantum Rabbit Laboratory?) and Quantum Rabbit founder Edward Esko.

Esko: What if the new theory of element formation wasn't a theory at all, but an entirely new paradigm, a paradigm shift, so to speak? What I'm proposing may be an entirely new context that encompasses the existing theories but from a new angle.

The new angle applies as much to elemental, atomic, nuclear, and subnuclear forces as it does to gravity and the movement of the stars and planets. Within this new cosmology, the varied theories of science are embraced and explained anew. The key is the origin of the force holding the nucleus together and the force that makes objects fall to the surface of the planet. My view is that this force (which takes many forms, including magnetism, electricity, gravity, etc.) doesn't originate from the body in question, be it a nucleus or a planet, but from infinite space. It is not a pulling force but an infinitely pushing force.

Grabiak: Paradigm shifts are very common in physics, and they usually break new ground without pushing aside the existing theories completely. Newton's laws of mechanics are still extremely useful, even after Einstein's theories of relativity and quantum mechanics have proven it wrong. This is usually a question of what scenarios you are dealing with. Relativity comes into play when dealing with objects moving close to the speed of light, and quantum mechanics when dealing with objects at atomic distances, for anything within our daily experience Newton's law of mechanics work quite excellently. The problem I see, if you are encompassing the existing theories, it would imply that the conclusions reached from those theories are still valid. Among other things this means that the Coulomb barrier is still there, thus the new paradigm does not help to provide a new explanation for LENR.

Esko: Let me try to sum up:

1. The attractive force between opposite charges is generated from the outside, so that they are being pushed together like opposite magnets.
2. Elements placed between the two charges will undergo changes, depending upon the degree of charge and other conditions, for example, pressure; vacuum versus atmosphere.
3. Lighter elements may fuse to form heavier elements; heavier elements may fission to form lighter elements, at least on the nano-scale.
4. The tendency of all matter is to condense into an infinitesimal point due to the force generated by the surrounding quantum vacuum. The reason this doesn't occur is because of the opposite, electrical repulsive force that exists between like charges, for example, proton and proton.
5. By manipulating pressure toward the quantum vacuum, and altering, even slightly, the plus or minus charge of two nuclei relative to each other, it may be possible to overcome the electrical repulsive force to the extent of allowing a small percentage of nuclei to bunch together to form a heavier nucleus. The quantum vacuum holds the nuclear particles of the new heavier nucleus together.
6. This process may also be occurring on the macrocosmic scale at the level of galaxies. Galaxies are the matrices in which matter is formed in the universe.

7. Observing the galaxy from the side, the force converging toward the center from the left represents one of the charges (anode) and the force converging from the right, the other charge (cathode.) Hydrogen and helium are continuously being formed from the condensation of energy into electrons and protons. These elements make up the primordial gaseous cloud that gives rise to the galaxy. These elements are squeezed toward the center by the quantum vacuum and, once they arrive there, are energized by the two oppositely charged vectors, appearing as two converging lines of force.

It is here that these light elements are forced outward toward the periphery, while undergoing nuclear transmutation into the heavier elements. As nuclear and elemental material is ejected from the center of the galaxy toward the outer spiral orbits, billions of tiny condensations occur in which the process of nuclear transmutation continues through the periodic table, arriving at the radioactive elements. These tiny condensations represent stars and solar systems. This process is taking place throughout the universe and has no fixed origin from the macrocosmic perspective. Each galaxy, however, has a defined beginning and end point that can be measured from our relative perspective. If you look at a photo of a QR vacuum tube with element plasma inside, it bears a striking resemblance to the galaxy side view. Perhaps we are, on the microscopic level, replicating a process that takes place continually on the galactic plane. This new paradigm replaces the big bang and other creationist interpretations of the genesis of the universe, including non-scientific interpretations.

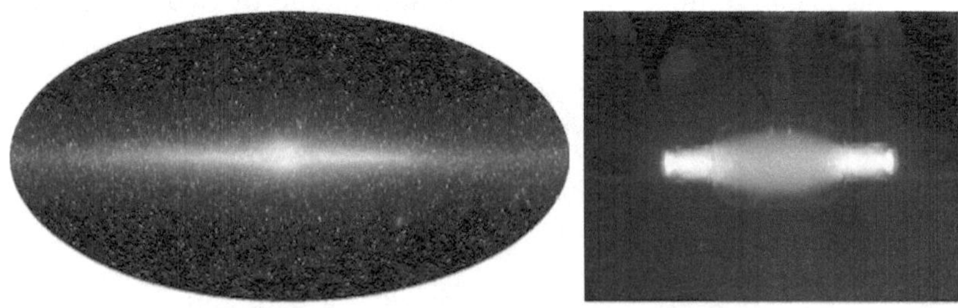

Left: Side view of the Milky Way Galaxy. Right: Helium plasma in QR vacuum tube

Grabiak: I would still like to stress that with point 5, you are in complete contradiction with the well-established laws of electromagnetism, postulating a radical change like that means that your theory will likely make lots of other predictions that contradict existing evidence, likely untangling your theory before it even comes together. It is absolutely not as you suggested that your theory encompasses the existing theories, it is going full out on a collision course.

Question, is it right that the universe knows to push opposite charges together and like charges apart? Can you provide any details of how that works? I have a few more thoughts. How about Einstein's two equivalent viewpoints of gravity? When observing a falling elevator from outside we would say the objects inside are accelerated towards earth by gravitational forces, but an observer within the elevator that is accelerated at the same rate as the objects inside would see the elevator as a chamber with no gravitational field inside and no forces acting on the objects. Can you accommodate that within your viewpoint? Is the universe pushing or is it not pushing? That example can also be turned around. Let's say there is an elevator that has been abducted into outer space, and an alien being starts pulling it and accelerating it.

From the inside of the elevator it looks like a gravitational force with objects appearing to fall in the direction away from the pulling alien, but an outside observer would say that there is no force at all acting on the objects inside. Again, is the universe pushing the objects or is it not pushing? Question, why has the concept of local field theories been so successful guiding us towards theories that have been very successful explaining nature? All current theories are of that nature, gravity, electromagnetism, and strong and weak interactions.

The idea is that the behavior of particles and fields is only determined by their immediate surroundings. An electric particle reacts to the electric field at its position, no matter if that field is caused by a charged particle on one side, or an oppositely charged particle on the opposite side, whether the field is caused by a small charge nearby or a big charge far away. This appears to be in stark contrast to your ideas. So, if your ideas are correct, can you think of a way of showing that local field theories are wrong, and a faraway charge in the universe does have an effect? (Side question, how would the effect of the universe be communicated? The concept of local field theory works well with the idea of "signals" (e.g. field waves) that travel at finite speed). On the other hand, if you were encompassing the existing theories, you would have to explain why the paradigm of local field theories is not a contradiction to your paradigm.

Esko: In the final analysis, it's premature for us to be speculating as to a theory to explain the QR low energy transmutation results. Most observers wouldn't accept the QR data to begin with. Our methods and controls were admittedly impromptu, makeshift, and not well controlled. The key objection—that of contamination beforehand has not been adequately addressed. We have, for example, been using the Certificates of Analysis provided by the supplier for the control values of the various elements used in the tests, rather than actual analysis of the control samples beforehand. Contamination could have crept into our results from myriad angles, from the handling of the materials prior to and after the tests, to the composition of the electrodes and glass tube materials, to the air quality in the room.

Given our level of sophistication, we can't say for certain whether or not low energy transmutation is in fact taking place, or is in fact real. However, my associates and I have seen results that seem to indicate that something unexplained is taking place in these tests. That something has occurred with such regularity and predictability that it strongly suggests our theory of low energy transmutation is valid and merits further testing under stricter controls. My point is that our evidence is not solid enough to challenge the current scientific paradigm. I would feel more comfortable if a university such as MIT or Berkley, or an established lab such as Cavendish or Oak Ridge, ran dozens of tests that validated our results. In other words, until low energy transmutation is established as fact, without a doubt, it is premature to debate the how or why it is or is not taking place.

If it turns out that low energy transmutation is confirmed beyond reasonable doubt, the great theorists of that time will do what they have always done, improvise a theory to fit the apparent facts. This may or may not occur during our lifetimes. I'd like to propose that we hold off on trying to come up with a theory for our results until more evidence comes in. The evidence will either add more proof to the existing paradigm, or force some type of revision or modification in it. I suggest we instead try to garner support, both scientific and financial, for another round of tests, using the published QR reports as the starting point.

Source: Correspondence between the author and Matthias Grabiak, 2013.

4

THE *INFINITE ENERGY* INTERVIEW WITH EDWARD ESKO

Briefly summarize the most profound transmutation results you have directly produced.

Our results are summarized in our books, *Corking the Nuclear Genie* and *Cool Fusion*. One powerful result was the apparent transmutation of zinc and sulfur into palladium under vacuum and at relatively low temperature. About 50 to 60 amps of electric power were delivered to the vacuum tube. A significant amount of chromium and strontium also appeared. Another dramatic result was the unexpected appearance of 1,500-ppm copper from the low energy transmutation of iron and lithium. In another test, 3,500-ppm germanium appeared from the transmutation of lithium and copper.

When you first found transmutation in an experiment, were the experiments aimed at transmutation or was it a byproduct of another process?

In our experiments we set out to prove the transmutation hypothesis. Transmutation wasn't a byproduct; it was what we were looking for.

What transmutation results (other than your own) seem the strongest?

Our work was inspired by the work of George Ohsawa, which was in turn inspired by the discoveries of Louis Kervran. Michio Kushi, a student of Ohsawa's, achieved dramatic results in experiments conducted in Cambridge in the mid-1960s. It was Kushi's lectures on this topic that directly inspired our work. The 19th century experiments conducted by Sir Norman Lockyer, the discoverer of the element helium and founder of the journal *Nature,* have also provided inspiration.

Do you think it is possible to transmute lead, uranium, thorium, etc.?

Yes. In our book, *Corking the Nuclear Genie*, I outline experiments for the low energy transmutation of uranium-235 and plutonium-239. If my hypothesis can be proven, it could solve the problem of nuclear waste.

Are there geological transmutations (apart from radioactivity)?

One of our colleagues, who is from MIT, hinted to me that transmutations may be occurring deep inside the earth. Transmutation may explain the continual appearance of certain metals with other metals in common ores. It may also explain the relative abundance of iron in the earth's crust. Our experiments with converting carbon to iron may hold the key. When lightning strikes a tree, for example, a portion of the tree's carbon may be converted into iron. Lightning may also cause two atoms of oxygen in the atmosphere to fuse, forming an atom of sulfur. My own feeling is that transmutation occurs constantly throughout the universe within giant electrically charged plasma clouds. If proven, such transmutations occurring at the galactic and intergalactic level would cast serious doubt on the Big Bang.

In the near future will it be possible to economically make noble metals by transmutation?

It has already been done. However whether small tabletop experiments can be scaled into large industrial processes remains to be seen. In the palladium experiment, for example, two inexpensive elements—zinc and sulfur—are used to produce palladium, a much more expensive element. If the experiment can be scaled, it could be quite profitable.

What benefits can society reap from the transmutation of elements in particular?

Transmutation of the elements is one side of the coin that will free humanity from the twin scourges of poverty and ignorance that have plagued it from the beginning. The other side is energy. Solving the mystery of the Great Pyramid, for example, may help greatly in that regard. For that we need to study the phenomenon of buoyancy, or as Edgar Cayce put it, "the way that iron swims" in water. An ocean of air surrounds us. It might be possible to apply the principle of buoyancy to float heavy objects, like the giant blocks of the pyramid, in the air. The principles of transmutation and unlimited energy are the principles of the universe. Understanding them will free humanity once and for all from the scourge of ignorance and the scourge of poverty.

Source: *Infinite Energy Magazine*, November-December, 2018.

Appendix

Biological Transmutation of Elements (1961)

Louis Kervran

We have repeated many times: the field of physics is widening. Things are changing in the world of physics, things are happening. When one follows recent research, often one has the impression that a revolution is in the making, a revolution comparable to the one started sixty years ago by the involuntary coalition of Einstein and Planck.

Just as the most intensely cold moment of the night announces the sunrise, one wonders if the increasing number of inexplicable facts theoreticians are now struggling with is not the prelude to the appearance of a new theory, a theory as important as that of relativity that will enlighten our present doubts and uncertainties with great and sudden light! Here are some of these facts, still largely unknown to the public.

A few months ago Canadian physicists discovered that atomic nuclei of two different elements could "encroach one another" (unite in a way that does not correspond to chemical combination, which is an electric, not a nuclear, phenomenon) and produce a new, stable nucleus. This "joining" phenomenon lasts something like a billionth of a second.

To what nuclear interaction does this phenomenon, which the Canadians call, "molecular effect of the nucleus," correspond? One does not know. According to a Japanese physicist whose works have just been published, mesons and hyperons would be true universes in ultra rapid evolution, so quick that the interactions that develop inside them operate at a speed greater than that of light. These mesons and hyperons (which are rightly called "strange particles") would evolve in another dimension that is neither time nor space!

But the strangest of all the facts coming to light in recent years was not identified by physicists. It is rather the chemists, or more precisely, the biochemists, who must receive the credit. We refer to nuclear metabolism, or the transmutation of the simple elements, one into another, by living beings.

In the April 1959 issue of *Science et vie*, we told of the experiments conducted at the Ecole Polytechnique by Professor Baranger. But now there is something else new. The experiments and observations we are going to discuss here confirm the phenomena already observed by Dr. Baranger on a humble grain, La vesce de Cerdagne. As Baranger anticipated, these biological phenomena should be universal: they should be verified in every living thing, and played an enormous part in the history of the earth.

EXPERIMENT IN THE SAHARA

From April to September 1958, a vast biological inquiry was conducted in the Sahara Desert, initiated by the Center of Research (PROHUSA, Human Problems in the Sahara). Its object was to study the behavior of the personnel engaged in the search for oil the Ouagla-Hassi Messaud-El Gassi region.

Very strict control was established on the food and drink intake of certain groups of workers. At the same time, excreta (stool, urine, and in some cases, sweat) was shipped by plane in refrigerated containers to the laboratories of the Faculty of Pharmacy at Strasbourg, where a team directed by Professor Hasselman proceeded to analyze it. Professor Metz, of the Faculty of Medicine, tabulated the results.

A year later, in April 1959, (at the time our study on the work of Professor Baranger was published), the Ministry of the Sahara sent a mission into the same desert region. I was appointed head of the mission. (Translator's Note: Kervran, an engineer, biologist, member of the Council for Hygiene for the Seine, and member of several technical interministerial commissions, is the author of classical works on the resistance of the human body to electricity. He has been interested in the problems of metabolism, i.e., the process of exchanges in living beings, for many years.)

My attention was attracted by the behavior of the personnel who were doing hard labor in the hot desert sun, on a metallic platform. In that area the outdoor temperature in the shade is higher than that of the human body. But the quantity of sweat evaporated (calculated by obtaining the difference between intake of food and drink and excretion by stool and urine) was only four or five liters per day. A simple calculation will show that, under these conditions, evaporation of sweat could in no way insure the constancy of temperature of the body, that is, 37 degrees C.

Hyperthermia never occurred. The temperature of these men should have gone up; they should have had immediate symptoms preventing them from working. But they had no trouble. Their temperature did not rise. Something was going on within them which absorbed excess heat, caused by the high external temperature and their work. What? That was what I wanted to know.

To solve the mystery, I again made a study of the results obtained the year before by PROHUSA. First I established the accuracy of data relative to the food and drink intake on the one hand, and to excreta on the other. I noted the length of the control (six months), which excluded any error due to progressive accumulation, or vice versa, to mobilization of elements taken from body reserves. Then I examined the tables themselves. Several of them (those concerning sodium, magnesium, potassium, and chloride) prompted my curiosity, for they were unexpected, to say the least.

1. The quantity of potassium excreted by perspiration in the form of potassium chloride is increased by high temperatures, and as a consequence, the proportion of potassium and chloride thus excreted increases, in comparison to the corresponding excretions of sodium. *All this was going on as if part of the sodium contained in the salt (sodium chloride) ingested had been transformed into potassium while passing through the body, but by what strange mechanism?*

   I remembered then the experiments of the American, Bass, in 1953. He too had collected human sweat precipitated by high heat, and observed that the body under such conditions excreted more potassium than it took in. No explanation could be found for such a surplus of potassium, which in the Sahara would rise to 50 percent in July.

2. In 80 percent of the cases, the bodily excretion of magnesium was markedly higher than the intake. And this finding was based only on the content of the urine and the stool (adding the sweat would only have increased this unexplainable difference). This surplus remained constant during the six months of observation, which excludes the hypothesis of using up the body's reserves. *Where, then, did the surplus of magnesium come from?*

3. Comparing the tables of calcium and phosphorus with those of magnesium, I could see that the two former were linked to the third. When the outcome was negative (that is, when the excretion of calcium was higher than the intake), so was the table for phosphorus; and each of the negative cases would correspondingly turn out negative for magnesium. Conversely, when magnesium was positive (in 20 percent of the cases), calcium and phosphorus were also positive.

All these phenomena suggest the same interpretation: If the body is capable of excreting chemical elements which it does not absorb, or if the body excretes more than it absorbs, and this continues for so long a period that any theory of depletion of the stock of body reserves is out of the question, one has to admit, since nothing is created, that these elements come from other elements! This seems quite mad, and contrary to the most sacred principles of chemistry and physics. But what can we do? Nothing is sacred in science, except fidelity to observation and experience. The actually admitted laws in physics and chemistry may have exceptions in the field of biology, exceptions that we did not see until now. I will show you more of these exceptions which defy any classical interpretation.

When a Hen Lays an Egg

In granitic country, where you find the elements of granitic desegregation, men and animals have a calcareous skeleton; birds lay eggs with calcareous shells. Yet there is no calcium (calcareous element) in the soil. In regions of granite origin, where the soil has long been eroded and only silicon and lime remain, one can observe decalcification of living beings. Why?

I had noticed that hens were picking small mica pieces all day long (mica is a silicate aluminum composed of potassium, iron, and magnesium). Could these hens distinguish mica containing calcium from mica not containing it? I made the experiment related below to find out.

I put laying hens in a hut that was built on sand made of clay and a silica (SiO_2), and fed them no calcium until they laid a soft-shelled egg. After this happened, I gave them pieces of mica. These hens never saw mica before, but they ate it as soon as they saw it. The next day they laid hard-shelled eggs.

Analysis of this experiment was very interesting. The mica contained Si, Al, K, Fe, and O, in 98 percent. The rest was Ti, Mg, and Ca. The proportion of mica was 200g to .003g Ca. However, the eggshell had 7g of Ca. Where did this 7g come from? Neither Si, Al, nor Fe diminished; not from O in the air. The Ca must have come from K only.

Then we have to ask ourselves whether K changed to Ca. Or did K draw Ca from the bones? If the latter, how can a hen continue to so draw from her bones all her life? Or does Ca come from other sources? Some may say that K works as a catalyst in foods. But there was not enough Ca in the hens' food to make a 7g eggshell.

GENERAL PRINCIPLES

If these experiments are accurate, how can we explain them? There must be metabolism of atomic nuclei. This is not a process of atoms or molecules, but a biological process in the nucleus. However, I have no intention of propounding a new theory of nuclear structure.

The most important thing is the fact that the biological transmutation of elements, which has been observed and reported by Messrs. Vogel, von Herzeele, Baranger, and I, can be confirmed by simple calculation: the addition or reduction of the oxygen or hydrogen nucleus. It is fortunate that these experiments and observations follow the same principle.

Case 1: As shown in the PROHUSA reports, and Mr. Bass in 1953, the fact that sodium (Na) disappears, and potassium (K) appears, can be explained in the following equation:

$$^{39}_{19}K - {}^{23}_{11}Na \rightarrow {}^{16}_{8}X$$

X will be oxygen because oxygen has atomic weight 16, atomic number 8; then the above equation can be written:

$$K - Na \rightarrow O$$

Case 2: The problem between calcium (Ca) and magnesium (Mg), reported by the Pharmacie College of Strasborg, will be explained by the next equation:

$$^{23}_{11}Na + {}^{1}_{1}H \rightarrow {}^{24}_{12}Mg$$
That is to say, Na + H → Mg.

Case 3:
$$^{24}_{12}Mg + {}^{16}_{8}O \rightarrow {}^{40}_{20}Ca$$
This shows that Mg + O → Ca.

Case 4: A hen laying an egg produces Ca from somewhere, but not from food, mica, or her own bones. Thus, Ca can come only from potassium; and the equation is:

$$^{40}_{20}Ca - {}^{39}_{19}K \rightarrow {}^{1}_{1}X$$
This X should be H.
Therefore, Ca - K → H.

METABOLISM OF THE NUCLEUS IN NATURE
If this transmutation in the biological field exists, it should exist also in the fields of astronomy and geology. The question arises whether a big geographical layer is produced by this kind of force or not? Whether it is a trace

of this kind of force or not? It is really so. We have found proof of it. First of all, there is the nitre. For several centuries nitre was the only natural source of nitrogen potassium (NK). NK exists in the hot or wet wall or in the lime sand produced in countries where wet and hot seasons come one after another.

The amount of the nitre grows by the action of bacteria at the beginning of a dry season. Silica is a nitrogen compound of lime, caustic potash, and magnesium oxide. However, how or from where do potassium and magnesium come in the pure lime sand? There is Ca at first, then K (potassium) and Mg (magnesium) appear by the action of the bacteria. This is shown in the equation below:

$$Ca - H \rightarrow K$$
$$Ca - O \rightarrow Mg$$

Should we accept that bacteria acts like a chemical agent which picks out hydrogen and oxygen from Ca (calcium)? We can't explain this yet at our stage of science; therefore, we must accept this equation.

Here is another example. Magnesium increases in dolomite when it oxidizes. Where does this magnesium come from? Someone said this is a meta-somatose, but this is not a solution for the question; only an exchange of unknown words. However, this phenomena can be explained by the same equation again: $Ca - O \rightarrow Mg$.

How wonderful this enigma of the dolomite is can be explained by the equation we get from biological observation.

There is a more interesting example which is accepted popularly. A bacteria works as an agent to produce petroleum. It is accepted also that the development of the layer of petroleum is related to the magnesium which exists in the dolomite. Then we get the equation:

$$Mg \rightarrow {}_2C$$

This equation, plus the previous equation ($Ca - O \rightarrow Mg$), leads us to the equation of petroleum. From this equation we can see that the source of petroleum is lime rock or sand.

$$Ca - O \rightarrow Mg \rightarrow {}_2C$$

This equation leads us to a reconsideration of research. The dolomite is an important point. The similarity between this equation and the real condition of the geographical layer has been proved by many geologists.

Another example, silicon (Si) and aluminum (Al) have a close relation in nature:

$$Si - H \rightarrow Al$$

This happens only on the surface of the earth, but the amount of aluminum in the earth decreases deeper in the earth. This can be explained by the fact that silicon becomes aluminum when the nucleus of silicon loses one nucleus of hydrogen, and aluminum becomes carbon when the nucleus of aluminum loses one nucleus of oxygen.

$$Al - O \rightarrow C$$

This coincides perfectly with geological observation.

Source: "Biological Transmutation of the Elements," by Louis Kervran, *Macrobiotic News*, May 10, 1961, Ohsawa Foundation, New York.

APPENDIX

*The New Discovery of the
Transmutation of the Atom (1965)*

Michio Kushi

Over the past several years, it has been discovered that the transmutation of the atom can be made, and is constantly going on, in biological bodies and in nature, under normal atmospheric conditions, with low temperature, low pressure, and low energy. This discovery has not only been observed in biological and natural phenomena, but has been duplicated experimentally by groups of scientists in Japan and France.

This discovery is a revolutionary event in the study of nuclear science, and it will bring vast influences to modem thought and civilization. These influences are so great that we cannot see at present the total effects and results that will come from the application of this discovery. Following are only a few of the changes we can expect in the future:

1. Chemistry and physics shall change their conventional theories due to the fact that all atoms are constantly changing from one to another, under normal atmospheric conditions, with low temperature, low pressure, and low energy.
2. The understanding of biology, biochemistry, medicine, and nutrition shall be greatly improved with the knowledge that all biological bodies, including the human body, are transmuting the atom within their bodies with low temperature, low pressure, and low energy.
3. Astronomy, geology, and other natural sciences shall gain a wider and more realistic understanding of the origin, process, development

and future of their fields with the knowledge of the real nature of atomic structure and the laws of its transmutation.

4. Industry shall begin to apply the transmutation of the atom to produce all elements and materials necessary for the development of civilization at a very low cost, with a simple process, in unlimited quantity, from several light elements easily obtainable from air, water, and soil.

5. The industrial revolution shall change the modem social structure, which may create the possibility of establishing a commonwealth of humanity, resulting in the opening of a golden age of prosperity and wisdom. This shall also eliminate ideological conflicts in economy and politics, which will serve for the avoidance of world nuclear war.

A Short History

Louis Kervran, a French scientist, introduced his discoveries and theories of the transmutation of the atom in 1960 through a scientific journal. Since then, he has presented his discoveries through writings and lectures. He has also published two books, *Transmutations Biologique* (1962), and *Transmutations Naturelles* (1963). His observations and theories have gained agreement and support from other scientists in France. His discoveries have also been confirmed by scientists and people in other countries. He and his scientific associates are continuing their study of the transmutation of the atom in France.

Among those who have encouraged Louis Kervran for this period, there is an outstanding Japanese philosopher, Yukikazu Sakurazawa, widely known as George Ohsawa in America and Europe, for his fifty years of devotion to the biological and physiological recovery of humanity. He has not only inspired Kervran, but has also introduced the transmutation of the atom to other countries, including Japan.

George Ohsawa and his French and Japanese associates started their laboratory experiments on the transmutation of the atom with low temperature, low pressure, and low energy early in 1964. Their success is based not only upon the discoveries made by Kervran and other scientists, but also on the traditional Oriental philosophy, namely the principles of change, or yin and yang, which are applied through modem scientific concepts. In accordance with the principles of yin and yang, everything in this universe is constantly changing, including the structure of the atom. The constant change also contains the key to explaining the cosmological development from pure energy to the world of atoms, then, to the world of chlorophyll, then to hemoglobin and man.

George Ohsawa's experiments have achieved important transmutations under normal atmospheric conditions, with low temperature, low pressure, and low energy. The results have been examined and confirmed by authoritative testing agencies in Japan and France. The experiments have achieved both physical and biological transmutations. By their experiments, they have discovered the unknown nature of the atom.

They have further studied methods for applying the transmutation of the atom in the actual production of elements and materials, which will produce them at a low cost, using several light elements. Patents have been filed on these methods, and further technical and engineering studies are being conducted.

DISCOVERIES

The successful experiments on atomic transmutation have created new knowledge of the atom since June 1964. These discoveries will be more extensively and scientifically published at a later time. Below is an outline of these discoveries.

1. All heavy elements have been produced on the earth by the transmutation of the atom through a process of nuclear fusion between light atoms (the elements from atomic number 1 through 8).
2. Those transmutations producing heavy atoms have been made by either physical transmutation or biological transmutation.
3. Those transmutations have been achieved and are being achieved under normal atmospheric conditions with low temperature, low pressure, and low energy.
4. Those transmutations can also be achieved by modem technological

and engineering methods, which shall achieve unlimited production of elements at an extensively low cost, because of the unlimited supply of light atoms in nature.

5. Those transmutations are conducted in accordance with the laws of natural change, conforming to modem scientific interpretation, and seen with great insight by ancient Oriental principles.

6. Applied transmutation not only produces the presently known elements, but also may produce unknown elements, as well as an unknown quality of materials.

7. Nearly all of the metallic elements have been transmuted from the lighter elements within a natural or rather cold atmospheric condition.

8. Atomic transmutation is made not only from lighter elements to heavier elements, but also from heavier elements to lighter elements, in accordance with changes in external conditions.

There are further discoveries, unknown to modem atomic theory, which explain the genesis and development of the universe.

Experiments

Following is a brief outline of experiments conducted in Japan and France. Some of these experiments are physical transmutations and others are biological transmutations. These are only examples, and complete information will become available in the near future.

1. The Transmutation from Na (Sodium) to K (Potassium).

The applied formula: $^{23}_{11}\mathrm{Na} + ^{16}_{8}\mathrm{O} \rightarrow ^{39}_{19}\mathrm{K}$

George Ohsawa and Masashiro Torii, Professor of the Musashino Institute of Technology, with several scientists, have achieved this transmutation of the atom from Na to K by the following method, first achieved on June 21,1964.

In this experiment, one electric discharge vacuum tube with two poles was used. The length of the vacuum tube was 20 cm and the diameter 2 cm. Electric poles of several different metals were tested. The power of electricity used in this experiment was 60 watts. First, 2.3 mg of Na was inserted and sealed in the vacuum tube, and electricity was started running through the tube. About thirty minutes later, 1.6 mg of O was introduced, and a second later, Na changed into K. This result was examined carefully by authoritative testing agencies, and the same experiments were performed repeatedly, yielding the same results.

2. The Production of Fe (Iron) from C (Carbon) and O (Oxygen).

The applied formula:

$$_2(^{12}_{6}C + {}^{16}_{8}O) \rightarrow ({}^{28}_{214}Si\ {}^{56}_{28}X) \rightarrow {}^{56}_{26}Fe\ \text{(+ 2 protons)}$$

George Ohsawa and his associates in Japan succeeded in their experiments with several methods to produce Fe from C and O. Later French scientists tested similar methods and confirmed the success of the transmutation. After creating the method to achieve the most efficient possible transmutation, they filed patents accordingly. The following examples show only a few methods to accomplish the transmutation from C and O to Fe.

Method 1: Transmutation in Air (A):
Two graphite crucibles (approximately 2.5 x 5 to 6 inches) cover each other top and bottom. The upper crucible has a 10 mm hole, surrounded by a ceramic ring. The ceramic ring acts as an insulator. Into this hole, a carbon rod (0.25 inches in diameter) is inserted until it reaches to the carbon powder (2 to 3 grams) placed at the inside bottom of the lower crucible. The lower crucible has one or two small holes at the lower part of its side wall for air circulation. An iron base placed under the lower crucible acts as one electrode pole. The carbon rod acts as another electrode pole. As the carbon rod approaches the carbon powder, an electric arc arises. Continuing the operation for 20 to 30 minutes, the carbon powder changes to Fe. In this experiment, the applied electricity is about 35 to 50 volts, and 8 to 18 amps, either AC or DC.

Method 2: Transmutation in Water:
Using two carbon rods (0.25 inches in diameter), create an electric arc between them, by striking them on one another in water. This operation is performed for 1 to 5 seconds. Then, brown-black metallic powder falls down to the bottom of the water, which contains Fe. The applied electricity is the same as in Method 1.

Method 3: Transmutation in Air (B):
Carbon powder is placed on a copper plate, approximately 12 inches long, 6 inches wide, and 0.5 inches thick. This plate works as an electrical ground. A carbon rod (identical to the carbon rods used in Methods 1 and 2) used as another electrical pole, strikes repeatedly the carbon powder on the plate, producing an electric arc. The carbon powder changes into Fe. The applied electricity is the same as in the above methods.

During the process of transmutation, Ni (nickel, shown in the formula as X), is temporarily produced. But it disappears very soon, for it is an isotope with a radioactive nature. The life of an Ni isotope is considered approximately 1/1000th of a second.

In these experiments, the degree of transmutation from C and O to Fe is approximately 5 percent to 20 percent immediately, with a larger percentage of transmutation occurring gradually in the air, which has the effect of cooling the metallic powder to below room temperature. The Fe that is produced by this transmutation is stainless. It does not rust easily. It has also much less reaction to heat than ordinary iron, due to its composition of 2 Si (silicon) as the formula indicates. This iron was named GOS. (George Ohsawa Steel), given the initials of George Ohsawa by the scientists who worked with this transmutation.

All results of the transmutation of Fe have been carefully examined and analyzed by several methods including: magnetic inspection, spectroscopic analysis, chemical analysis, and examination by reagent, confirmed by authoritative testing agencies.

3. Biological Transmutation of the Atom
George Ohsawa has guided Sanehide Komaki, a professor of biology at Mukogawa University, in Kyoto, Japan, and has accomplished several

methods of the transmutation of the atom through biological cells, using bacteria or microorganisms. Among these transmutations, some applications are creating results that could prove of tremendous benefit to humanity. Similar experiments have been conducted in France by Louis Kervran and other scientists, which are also producing noteworthy results. Examples are outlined below:

Experiments with bacteria and microbes:
The ferment of beer, black mold, or a similar microbe, which has been cultivated without K (potassium), grows and increases, first slowly but later rapidly, producing K within itself, by the transmutation from other elements. With a method applying this principle, K grows, increasing up to 100 times in quantity after only 3 days; 10,000 times after only 6 days; 10^{20} times after 30 days, and so on. This is the method to produce and supply the total quantity of K that the entire world needs for agricultural and industrial use. Similar productivity was achieved when the s. cereisae dried microbe was cultivated on the non-K land.

According to calculations based on realistic and scientific cultivation, a land space of about 2,000 acres, having the proper storage and facilities, can produce 200,000 tons of K, and 20,000,000 tons of protein every year. The above amount of K becomes 240,000 tons of potassium fertilizer (K_2O), and the amount of protein is equivalent to 50,000,000 tons of beef. These amounts will supply an ample quantity of protein and potassium as the essential food supply for the entire society of mankind. (Detailed information concerning the above methods will be soon introduced in the academic report of the Biochemical Society, Paris, France.)

Experiments with vegetable seeds:
Vegetable seeds such as buckwheat, soybeans, peas, and others, which have been cultivated in twice distilled, purified water, begin to grow gradually but in perfect condition, with all elements within the seeds increased 100 to several hundred times in 30 days. This demonstrates that biological life is transmuting within itself elements suitable for its specific needs. These biological transmutations are going on within plants at atmospheric pressure, temperature, and with normal solar energy.

CONCLUSION

The above report presents only a simple glance at the problem relating to the transmutation of the atom. The purpose of this report is to give a basic understanding of the facts and possibilities of atomic transmutation with low temperature, low pressure, and low energy. In accordance with the analysis report from Japan, it was confirmed that more than 35 elements such as Si, Al, Ca, Mg, Mo, Ni, Na, K, Pt, Hg, Au, and so forth can be produced from the light elements (atomic numbers 1 through 8). The efficiency of production will depend on further technological and engineering refinements.

The Japanese scientists who worked on these initial experiments on atomic transmutation included:

George Ohsawa—Philosopher, writer, honorary professor at Nippon University, honorary citizen of Paris, founder and president of Institut des Hautes Etudes Dialectiques et Scientifiques, Tokyo.

Masashiro Torii—Doctor of chemistry, professor at Musashino Institute of Technology, Tokyo.

Shizuko Washio—Doctor of biology, professor at Ato- mi University, Tokyo.

Sanehide Komaki—Doctor of agriculture, professor at Mukogawa University, Kyoto.

Chikao Narita—Doctor of medicine, president of Tokyo Shibaura Hospital, Tokyo.

Yuzuru Sasaki—Research member of Institut des Hautes Etudes Dialectiques et Scientifiques, Tokyo. Noburu Yamamoto—Research member of Institut des Hautes Etudes Dialectiques et Scientifiques, Tokyo.

In addition, five universities and two other chemists cooperated in these studies and experiments. In France, thirty members organized a research group for the transmutation of the atom, including scientists, scholars, businessmen, and students.

Source: "The New Discovery of the Transmutation of the Atom," by Michio Kushi, *Macrobiotics Study Report,* Volume 1, July, 1965, East West Institute, Cambridge, Mass.

About the Authors

Michio Kushi, leader of the international macrobiotic community for over a half-century, introduced natural and organic foods, alternative and complementary medicine, and novel approaches to war and peace, energy and the environment, and other planetary issues.

Born in Japan on May 17, 1926, he studied international law at Tokyo University and came to the United States in 1949. Influenced by the macrobiotic teaching of George Ohsawa, with whom he studied in Japan and later on his visits to America, Michio and his wife, Aveline (1923-2001) settled in the Boston area. In the 1960s, the Kushis founded Erewhon, the pioneer natural foods company, and in the 1970s and 1980s founded the East West Foundation, *East West Journal,* Kushi Foundation, and the One Peaceful World Society. In 1985 he became General President of the World Federation of Natural Alternative Medicine, an association of 300 organizations.

Michio and Aveline traveled around the world, spreading macrobiotics and a natural way of life. They gave hundreds of seminars in Europe and

Japan, and Michio led a seminar for medical experts convened by WHO (World Health Organization) on AIDS and diet in Central Africa.

Michio inspired research on the macrobiotic approach to heart disease, cancer, diabetes, and other chronic disorders at Harvard Medical School, the Framingham Heart Study, the National Institutes of Health, Center for Disease Control and Prevention, and other universities, medical associations, and government agencies. The Smithsonian Institution created a permanent Kushi Family Collection on Macrobiotics and Alternative Health Care at the National Museum of American History in Washington, D.C. in 1999 and held a gala celebration in honor of him, his wife, and the macrobiotic community. The U.S. House of Representatives unanimously passed a resolution praising his contribution to the health and well being of modern society.

Over the years, Michio authored more than fifty books on personal and planetary health, including *The Book of Macrobiotics, The Cancer Prevention Diet, The Macrobiotic Way,* and *One Peaceful World.* He received the Award of Excellence from the United Nation's Writer's Society. Michio continued to teach and counsel actively until his passing on December 28, 2014 at age eighty-eight.

Edward Esko, is one of the world's most active contemporary macrobiotic teachers. Over the past four decades, he has lectured and counseled in Europe, Asia, Latin America, the Middle East, and throughout North America, including at the United Nations, and has written and edited numerous books and articles. Building on the teachings of George Ohsawa, Michio Kushi, and other macrobiotic pioneers, he has applied yin and yang—the universal principles of change and harmony—to helping solve issues of personal and planetary health. He has served as Executive Director of the East West Foundation and Director of Education at the Kushi Institute. He is the founder of the International Macrobiotic Institute (IMI) in Massachusetts, and serves on the Board of Planetary Health, Inc., which is the non-profit sponsor of the Macrobiotic Summer Conference. He is the founder of Quantum Rabbit, LLC and The Quantum Research Institute.

QUANTUM RESEARCH INSTITUTE

The Quantum Research Institute (QRI) was founded in 2018 by Edward Esko to further the work of Sir Norman Lockyer, Louis Kervran, George Ohsawa, Michio Kushi, and other pioneers in the field of low energy transmutation, including the ideas and experiments presented in this book. Networking with like-minded colleagues around the world, including Mahadeva Srinivasan (India), George Egly (Hungary), Vladimir Vitsotskii (Ukraine), Wal Thornhill (Australia) and others, QRI is actively seeking solutions to global problems such as the increasing scarcity of rare metals and the nonstop accumulation of nuclear waste. Contact QuantumResearchInstitue.com for information or email: edwardesko@gmail.com.

BOOKS ON LOW ENERGY TRANSMUTATION

Amberwaves Press, 2012
Paperback: 176 pages
ISBN-10: 1477563725
ISBN-13: 978-1477563724

Amberwaves Press, 2013
Paperback: 88 pages
ISBN-10: 1493664743
ISBN-13: 978-1493664740

IMI Press, 2018
Paperback: 39 pages
ISBN-10: 1729588735
ISBN-13: 978-1729588734

IMI Press, 2018
Paperback: 61 pages
ISBN-10: 1729780164
ISBN-13: 978-1729780169